CW01510509

SCIENCE AND MAGIC IN THE MODERN WORLD

Science and Magic in the Modern World is a unique text that explores the role of magical thinking in everyday life. It provides an excellent psychological look at the subconscious belief in magic in both popular culture and society, as well as experimental research that considers human consciousness as a derivative of belief in the supernatural, thus showing that our feelings, emotions, attitudes and other psychological processes follow the laws of magic.

This book synthesises the science of 'natural' phenomena and the magic of the 'supernatural' to present an interesting look at the juxtaposition of the inner and outer selves. Fusing research into psychological disorders, subconscious feelings, as well as the rising presence of artificial intelligence, this book demonstrates how an engagement with magical thinking can enhance one's creativity and cognitive skills.

Science and Magic in the Modern World is an invaluable resource for those studying consciousness, as well as those looking at the effect of magical thinking on religion, politics, science and society.

Eugene Subbotsky obtained a PhD in psychology at Moscow State University in 1975, taught Psychology in Russia and the UK and conducted research in Russia, Germany, UK, USA and Mexico. Most of his works are on magical thinking in modern humans, cognitive development, the development of metaphysical thinking and moral development. He has authored 14 books and over 120 research papers and is an associate fellow of the British Psychological Society, a BPS Charted Psychologist, a member of the BPS Division of Academics, Researchers and Teachers. He is currently a reader (emeritus) at Psychology Department, Lancaster University, UK.

SCIENCE AND MAGIC IN THE MODERN WORLD

Psychological Perspectives on Living with the Supernatural

Eugene Subbotsky

Routledge
Taylor & Francis Group

LONDON AND NEW YORK

First published 2019
by Routledge
2 Park Square, Milton Park, Abingdon, Oxon OX14 4RN

and by Routledge
711 Third Avenue, New York, NY 10017

Routledge is an imprint of the Taylor & Francis Group, an informa business

British Library Cataloguing-in-Publication Data
A catalogue record for this book is available from the British Library

Library of Congress Cataloging-in-Publication Data
Names: Eugene Subbotsky, author.
Title: Science and magic in the modern world: psychological perspectives on living with the supernatural / Eugene Subbotsky.
Description: New York : Taylor and Francis, 2019. |
Includes bibliographical references.
Identifiers: LCCN 2018018092 | ISBN 9781138591349 (hardback : alk. paper) | ISBN 9781138591455 (pbk. : alk. paper) | ISBN 9780429490378 (ebook)
Subjects: LCSH: Magical thinking. | Psychology–Research.
Classification: LCC BF1621 .S833 2019 | DDC 133.4/3–dc23
LC record available at https://lccn.loc.gov/2018018092

ISBN: 978-1-138-59134-9 (hbk)
ISBN: 978-1-138-59145-5 (pbk)
ISBN: 978-0-429-49037-8 (ebk)

Typeset in Bembo
by Out of House Publishing

CONTENTS

PREFACE

Acknowledging the existence of supernatural reality, which sometimes quite wrongly is identified with superstitions, is often considered a violation of the scientific view of the world. At the same time, few people would deny that the world we live in is bigger than the universe studied by physical sciences. Social sciences, such as economics and political sciences, deal with realities that do not conform to the universal and unchangeable laws of nature. This is even more the case when it comes to disciplines such as theology and art. Many of the greatest philosophers of all time – Thales of Miletus, Pythagoras, Socrates, Plato, Plotin, Thomas Aquinas, Descartes, Kant, Hegel, Schopenhauer and others – in one form or another admitted that something exists that goes beyond the physical universe studied by sciences. Nevertheless, studies of the supernatural reality thus far have mostly been confined to esoteric theories of magic and witchcraft that are either detached from or stand in a direct opposition to the scientific approach to the universe. But could it be the case that supernatural events are as real as scientific ones, and that an ordinary human person lives in a world in which natural and supernatural events are mixed and intertwined? Is it not possible that both natural and supernatural realities coexist in one mind and in one universe? Finally, could natural and supernatural realities cooperate with one another towards a common goal – to make the life of a human individual more meaningful and manageable? In this book, I will try to answer these questions in the light of recent psychological studies on magical thinking and magical beliefs in modern humans.

Having been involved in experimental psychology for over 40 years, I eventually came to the conclusion that psychological phenomena have little in common with the natural phenomena that are studied by physical sciences. Thus, in nature, every atom and every molecule is an exact copy of other atoms and molecules of the same kind. In contrast, *psychological phenomena are unique*: every sensation, perception, thought and feeling is impossible to reproduce, existing only once. It is only with a crude approximation that psychological events can be brought to a common denominator and thus squeezed into the 'psychological laws', which to some extent look similar to the laws of physics. Further, we know that natural phenomena comply with the law of physical causality, according to which the same causes always produce the same consequences. Unlike the reactions of physical objects, *a person's reactions to the same stimuli are always slightly different*. Physical causality is based on four known types of physical interaction: gravitation, electromagnetism and the strong and weak nuclear forces, whereas *psychological causality both inside the mind (e.g., associative thinking) and in interpersonal interactions is based on meanings*. A physical body is passive and totally determined by the laws of nature, while *a person is active, freely determines his or her actions and is able to generate new actions and thoughts 'from nothing'*.

But if the human mind and actions do not obey the laws of nature, then what kinds of laws do they obey? Pondering this question, I more often than not came to the conclusion that *our mind and actions follow the laws of magic*. Magic is a way of comprehending the world that historically and logically precedes science. For millennia, people tried to influence nature by magic. Reflecting upon feelings and thoughts of their own, people assumed that animals, plants, rivers and mountains too have similar feelings and thoughts. By magical spells and rituals, people tried to influence animals and the processes of nature. Science put an end to these illusions, but only in regard to inanimate objects, whereas the mind remains under the power of magical laws. As a result, *the life of modern people evolved as a complex reality in which the laws of magic and science are intertwined*. Even though many scientists in public deny that they believe in magic, in their private lives, the same scientists often follow the laws of magic. Even in the domain of physical reality, science has approached processes that can be described in terms of magic rather than in terms of physics. In other words, *in modern life, magic and science complement each other*. It is this wonderful complementarity phenomenon that I would like to discuss in this book.

In English and other languages, the term 'magic' (which I will be using interchangeably with the term 'supernatural') can have a variety of meanings. In order to make the discussion of the relationships between magic and science comprehensible, we need first to define the concept of magic (the supernatural) that will be used in this book, and then to conceptualise the ways magical events relate to scientific ones.

INTRODUCTION

Magic and science

Science deals with natural phenomena, and magic with supernatural ones. But what exactly makes a phenomenon 'natural' or 'supernatural'? Because the terms 'natural' and 'supernatural' are rather basic, they need to be deduced from some concept that is intuitively clear for most readers. I suggest that our intuitively clear sense of our Self is taken as the basis for the distinction between the natural and the supernatural. The 'sense of Self' is the fact that at this very moment I, (and hopefully the reader, too), am having the feeling of being a certain individual with a name, gender, family and national belonging.*

Along with our Self, we also clearly have the feeling of something that we usually call the internal and external worlds. The *internal world* encompasses our feelings and experiences, such as a headache, the feeling of love or subconscious processes, whereas the *external world* consists of phenomena and ideas that exist outside our Self (e.g., the window in front of us, the wind behind the window that moves the tree brunches, the stars in the sky or the knowledge of physical laws). The internal world is accessible only to our Self, whereas the external world we share with other people. Our Self and internal and external worlds are parts of a more general concept, which is our mind (consciousness). The mind (consciousness) therefore is a concept that creates a final border for any sensible analysis. The mind can no longer be reduced to a more general concept. All that exists outside the mind is what Immanuel Kant called 'Ding-an-sich' ('Thing-in-itself') – some kind of reality that a person can only know about through his or her mind (see Figure 0.1).

The relationship between our Self and the internal and external worlds is a complex one[4]. In both of these worlds, features can be distinguished that our Self can directly control, while other features are beyond our ability of control. For example,

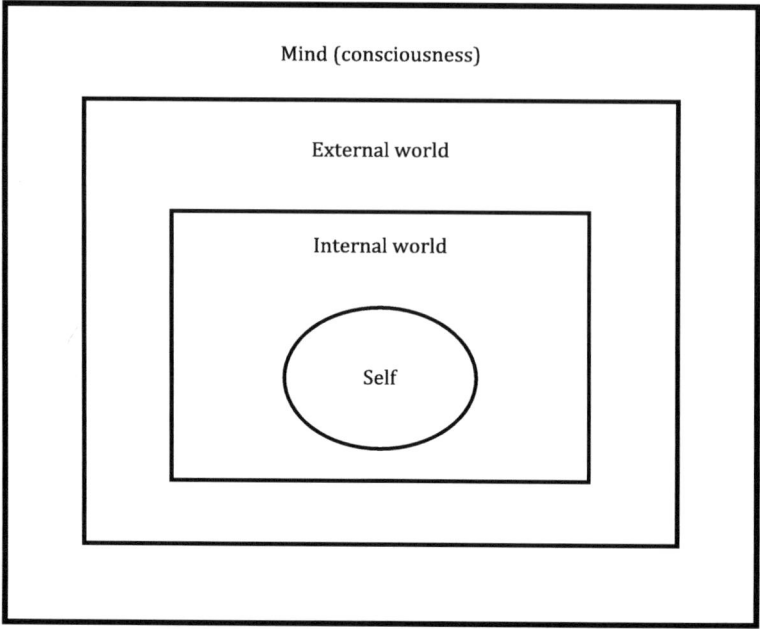

FIGURE 0.1   The structure of the mind (consciousness)

in the internal world we can control our attention (e.g., we can fix our gaze at will on this or that object, concentrate on some of our thoughts and personal beliefs). At the same time, we cannot stop our toothache by simply wishing it to stop, and we are unable to say the name of a famous actor that we know well but whose name at the moment is escaping our conscious effort to recall it.

In the external world there is only one object that our Self can control directly, and this object is our body. Thus, we can move our hand at will, but are unable to lift this computer by just thinking about lifting it. Let us call the objects that we can control directly (such as our attention, some of our thoughts or movements of our body) 'dependent objects', and the objects that we cannot directly control 'independent objects'. Usually, the class of independent objects includes all perceived external physical objects, such as inanimate things, animals and other people. It also includes mental objects: the laws of logic, scientific theories, myths, religions, social customs and social and moral laws. A subclass of inanimate independent physical objects is what we usually call *matter* and subclasses of independent mental objects are what we call *knowledge* and *collective beliefs*. Note that from this theoretical perspective *matter is a part of the mind (consciousness)*. Certain objects within our internal world, such as the feeling of pain or love, obsessive thoughts and subconscious processes, also belong to the class of independent objects (see Table 0.1).

TABLE 0.1 The relations between types of worlds and types of objects

Type of worlds Type of objects	Internal	External
Dependent on our Self	Our attention Some of our thoughts An effort of will	Voluntary movements of our body Certain patterns of electric impulses in the brain
Independent of our Self	Some of our feelings Obsessive thoughts Subconscious processes	External physical objects and processes (matter) External mental objects and processes (knowledge and collective beliefs)

It is also clear that we can act on physical objects with our body: for example, we can lift things with our hands and influence other people with our language. We can also act on mental representations of external objects (rational constructions) using the laws of logic and physics. Let us call this kind of action *indirect Self over matter/ mind actions*. Contrasting with indirect Self over matter/mind actions are our actions on our own body or mental processes; we will call this kind of action *direct Self over matter/mind actions*. Finally, external objects can act one upon another through physical forces (interaction in physics, chemistry and biology) or through language (social communication of people or certain kinds of animals) (see Table 0.2).

Now the distinction can be introduced between natural and magical interactions. *Natural interactions* include indirect actions of our Self on inanimate objects (such as a hand lifting a stone), our actions on mental representations of physical objects according to the laws of science and logic and interactions between inanimate objects, mediated by the known physical forces (e.g., wind moving a tree branch, the sun attracting the earth via gravitational force, a quantum particle colliding with another particle). Interactions between chemical substances, between bio- logical structures inside an organism and non-semantic interactions between living organisms belong to this class as well. In contrast, *supernatural (magical) interactions* are direct actions of our Self on inanimate physical objects (e.g., our thought affecting the muscles of our body or patterns of electical signals in the brain, a magic spell, a thought or a wish affecting a physical process), interactions between thoughts within our Self, non-semantic interactions between minds or between inanimate objects that are not mediated by any known physical force (e.g., extra- sensory perception or the entanglement effect in quantum physics) or interactions between nothingness and existence (e.g., emergence of random events or creative ideas).

Semantic interaction occupies the intermediate position between natural and supernatural interactions. Semantic interaction is carried out through physical

TABLE 0.2 Types of actions

Self on external and internal objects		External objects on other external objects	
Direct	Indirect	Physical	Social
Our voluntary actions on our own body	Our actions on external objects through our body, tools and language	Actions of one inanimate object on another inanimate object (e.g., a mechanical push, chemical reactions and interactions between structures of a living organism)	Interactions between people mediated by language, social customs and collective beliefs
Effects of our thoughts on patterns of electric activity in our brain and nerves	Our actions on mental representations of physical objects through the laws of logic and science	Certain interactions between animals (e.g., a predator catching its prey)	Social interactions between animals (e.g., between 'workers' and 'soldiers' in ant society)
Artistic and associative thinking			

phenomena – sounds or visual images – but it cannot be reduced to these phenomena. Thus, the meaning of the request "Would you please close the door?" cannot be deduced from the energy of sound waves spent on pronouncing and transferring this message. A receptor of this semantic communication performs the action of closing the door not because he or she hears the sound, but because he or she understands and approves the meaning of the message that the sound carries. Because the receptor of the message does not conform to the sheer physical force that the sound contains, but complies with the meaning that is carried by the sound, the semantic communicative interaction contains the element of the *direct Self over mind action*; in this sense, semantic communication can therefore be classified as a supernatural action. However, because semantic communication is carried out through physical forces, it is not a truly supernatural interaction. Rather, semantic interaction combines the elements of natural and supernatural interactions. We therefore may call semantic interactions *quasimagical interactions*.

By analogy with natural and supernatural interactions, one can talk about natural and supernatural phenomena. *Natural phenomena* are phenomena that are produced by natural interactions. Thus, crystals emerge from minerals dissolved in water as a result of their concentration, stars and planets derive from cosmic dust condensed by the force of gravitation, scientific theories result from manipulations with rational constructions of objects in accordance with the laws of logic and science and new species of animals develop through evolution. *Supernatural phenomena* are phenomena produced by magical interactions. Examples of such phenomena are changes in inanimate objects directly initiated by our mental processes and physical events that have no physical causes (e.g., a ghost, a random event or the emergence of the universe). Thus, according to the Book of Genesis, God created the earth and the sky from nothing. A creative thought in the human mind or the action of free will appear from nothing, too. A special magical phenomenon is *subjective experience* — sensations, perceptions, thoughts and feelings. As long as a person can access the world only through subjective reality, *subjective experience is a part of both natural and supernatural phenomena*. However, natural phenomena, through their rational constructions (knowledge and theories about these phenomena's structures, origins, functions and properties), are connected between themselves by causal links and form the continuum of *everyday reality*. Unlike natural phenomena, magical phenomena don't have a stable basis in rational constructions, are not connected between themselves and form *magical reality*. Because subjective experience is an underbelly of the mind (consciousness) as a whole, it will be considered separately (see Chapter 3).

Finally, *quasimagical phenomena* include such events as the perceptions of a piece of art by a person, myths, religions, social and moral laws (see Table 0.3).

To summarize, the following features distinguish magic from science. Magic is based on the supposition that a person and the things of nature are linked one to the other by some kind of 'prearranged harmony' or 'understanding'. In the magical world, the person can ask or order and the natural things can understand these requests and orders and comply or not comply with them (*direct Self over matter/ mind magic*); objects that resemble each other (e.g., a person and his or her photograph) are magically linked one with the other in such a way that if something happens to one of them (e.g., to the photograph), then the same happens to the other (e.g., to the person) at an indefinite distance and in no time (*sympathetic magic*); if an object is in physical contact with a person (e.g., a piece of clothing worn by a certain person), then this object acquires properties of the person (e.g., the good or bad features of the person's character) as if it is contaminated by some sort of 'magical infection' (*contagious magic*); and finally, invisible spirits can acquire physical form, as if emerging from nothing, and vanish again (*emergence/vanishing magic*). In other words, *in the magical world there is no watershed between objects that depend and don't depend on the human Self, and objects can interact one with another by transcending the limitations of space and time*. In contrast, the objects that science studies are supposed to be entirely independent from the human Self, and interactions between individual objects, such as galaxies or elementary particles, are constrained by physical limitations. These interactions need a certain medium (e.g., magnetic or gravitational fields) to conduct physical causation, require a certain time to proceed and the speed of physical interactions cannot exceed the speed of light (see Chapter 4 for more details on the differences between science and magic).

A brief look into history

The ancients believed that trees, rivers, the wind and other natural objects and events had minds and souls of their own, with which people could communicate

TABLE 0.3 Types of interactions and types of phenomena

Type of interaction or phenomenon	Examples of interactions	Examples of phenomena
Natural	A person affecting macro objects with hands, tools or devices	Formation of planets and stars from cosmic dust
	A person manipulating mental representations of objects via the laws of logic and sciencee	The growth of a crystal
		Formation of new chemical substances
	Electrical signals of the brain and nerves affecting the sensors of prosthetic devices	Evolution of new species of animals
	One physical object affecting another physical object via known physical forces	Manufacturing objects in industry or art
		Scientific theories
	Interactions between chemical substances, organs inside a body, between living organisms and between living organisms and the environment	
Quasimagical	Semantic communication between people	Perception of a piece of art by a person
		Myths, religions, social and moral laws
Supernatural (magical)	A person affecting inanimate objects by a spell or wish and without a physical medium	Objects emerging from nothing (e.g., the universe, a random event, a creative idea, an action of free will)
	Our Self intentionally moving our body	
	Our thoughts affecting patterns of electrical activity in our brain and nerves	Different objects sharing a common essence (a tribe and its totem, entangled quantum objects)
	Our thoughts, conscious and subconscious, directly affecting other thoughts by association	A conscious agent creating physical objects by an effort of will (God creating the earth)
	Our self affecting states of our body (placebo effect)	Subjective experience (sensations, perceptions, feelings and thoughts)
	Being in love with a person	
	A person passing information to other people without a physical medium (extrasensory perception)	

through magical spells and actions. Even in the Dark Ages this animistic belief was still alive. Following Aristotle, people of the Middle Ages believed that living creatures had souls, which made them similar to people. It was not until the middle of the seventeenth century that mechanistic science emerged in Europe. Rene Descartes and some other philosophers started to view animals as complex machines, similar to a mechanical clock. Eventually, the view of nature as a mechanical process was extended to humans as well. First, a human body started to be viewed as a mechanism, and then the human mind, too. A modern version of mechanistic psychology – cognitive science – understands the mind as a direct result of material processes in the brain. Phenomena that in earlier times were viewed as magical (e.g., heredity as a perpetuation of ideal forms of plants or animals, which existed in the supernatural world of 'ideas') today are explained by natural causes (e.g., heredity as the passing of information coded in the genes). Most educated adults today consciously believe in chemistry, astronomy and mathematics, and not in alchemy, astrology and numerology.

As a result of the progress of science, in the modern world reality is often identified with physical reality. But how valid is this identification? We know that the notion of physical reality itself grew out of myth and religion. Moreover, as argued above, physical reality is only a part of reality. The other parts are social and psychological realities: the realities of feelings, thoughts, fantasies, communication and language. These realities do not comply with the laws of science; rather, these realities now, as in ancient times, comply with the laws of magic. Movies and books for children are full of magical events, and young children's beliefs in Santa or the Tooth Fairy are even encouraged by adults. Our dreams and art often contain magical characters and events. Finally, in mental disorders, such as obsessive–compulsive disorder and schizophrenia, patients sometimes report experiencing supernatural images and events.

In other words, in the modern world science and magic coexist, and even cooperate one with the other in various domains of life. But how is magic built into the edifice of the modern world erected by science?

The complementarity principle

Sometimes a phenomenon can comply with two opposite rules or have features that contradict each other. In quantum physics, a combination of two opposite features in one object or event is named *the complementarity principle*.

For example, in the nineteenth century, light was viewed as a wave process, until in the beginning of the twentieth century Einstein suggested that light consists of particles – quants of energy. But how could light be a wave and a particle simultaneously, when properties of waves and particles are so different? Indeed, a particle has a certain localisation in space and a wave doesn't. A particle moves along a certain trajectory and a wave spreads in all directions simultaneously. Research has shown that in the micro world, units of matter and energy, such as photons, electrons and protons, do indeed combine the properties of waves and particles and, depending on the situation, behave either as waves or as particles. Scientists had no choice but to accept that the elementary units of matter have opposite properties simultaneously, and this gave birth to the complementarity principle[5].

In a similar vein, some phenomena can combine physical and magical properties. Take, for instance, the Big Bang – the event that gave birth to our universe. According to modern cosmology, the Big Bang is a physical event, which resulted in the emergence of the four main types of fundamental physical forces – electromagnetism, gravitation and the strong and weak nuclear forces. But at the same time, the Big Bang is a magical event, because at the beginning of the Big Bang, called the initial singularity, the universe must have had an infinite density at a finite time, when the fundamental laws of physics did not apply. In addition, space, time and causality emerged with the Big Bang: this means that the initial singularity itself emerged from nothing, and emergence from nothing is a characteristic of a magical event[6].

Another example of the complementarity between physical and magical properties is the effect of quantum entanglement[7]. According to this effect, if two elementary particles are generated in such a way that their total spin is zero, then a change in one of the particles (e.g., a change of the particle's impulse) is accompanied by a predictable change in the other particle at whatever distance the other particle is from the first particle. This means that the two particles are linked not through any of the four known fundamental physical interactions, which is physically impossible. Accordingly, the entanglement effect is a physical effect that also has the property of contagious magic – the ability of one object that once was in contact with another object to maintain this contact at an indefinite distance and in no time.

But perhaps the best example of complementarity is human consciousness. Every metal action is accompanied by physical processes, such as the transmission of electric impulses in neuronal synapses. However, for the person in whose brain the aforementioned physical processes unfold, these processes are represented not as physical events but as subjective experiences, such as sensations, dreams, feelings or thoughts. Subjective experiences cannot be described in terms of physical space. For example, if a person is experiencing pain, this feeling cannot be measured in metres, weighted in kilograms, described as a cuboid or a sphere, or presented in terms of electromagnetic or gravitational fields. If damage inflicted to the body is a physical, natural event, then the feeling of pain that accompanies the damage is a nonphysical, supernatural event. Paradoxical as it is, the reality of mundane, everyday subjective experience is a magical, supernatural reality (see Chapter 3 for more on subjective experience).

What this book is about[8,9]

This book's main aim is to discuss how magical reality is built into the reality of various domains of modern life. In Part I of the book (Chapters 1–3), issues regarding the role of magic in the human mind are raised. In Chapter 1, the argument is presented that art emerges as a means of early humans' communication with gods and spirits. Paintings on cave walls were the first symbols used to reach the domain of the supernatural. More abstract symbols of written language and mathematics eventually branched off from art, yet in its original and authentic form, art remains the pathway into the realm of the supernatural. Using paintings of surrealist artists, the chapter argues that these paintings address the implicit beliefs of rational people in magical reality. Harboured deeply in the subconscious, the belief of a modern person in the supernatural resonates with the paintings and elicits in the viewer feelings of mystery and anxiety, as well as insights into the meaning of life.

Chapter 2 is about individual consciousness. The hypothesis that consciousness is a uniquely human ability to simultaneously dwell in two types of reality is suggested: visible, ordinary reality and invisible, magical reality. This dual structure of consciousness makes humans capable of reflection, creativity and intentional action. But the same duality of consciousness also creates problems for human existence. One of these problems is the necessity to constantly make an effort to maintain the border between everyday and magical realities.

In Chapter 3, human consciousness is compared with artificial intelligence (AI). An argument is made that it is impossible to deduce subjective experiences from the processes in the brain; on the contrary, scientific theories about the structure of brain processes were initially derived from subjective experiences. This makes subjective experience a supernatural phenomenon. As long as only living entities possess subjective experiences, such experiences, as with life itself, cannot be artificially created. In contrast to subjective experience, AI is a product of programming. We can therefore conclude that AI, even in its most advanced forms, cannot simulate subjective experiences. This puts an impregnable barrier between AI and human intelligence. Neuroscience and cybernetics will create increasingly perfect interfaces between AI and human intelligence, but there will always be a 'neutral strip' of the unknown between 'intelligent machines' and human consciousness.

Part II of the book (Chapters 4–6) is about the relationships between magic, religion and science. In Chapter 4, the psychological phenomenon of a merger between magical and scientific thinking is analysed. The analysis reveals that some phenomena and theories in physics and cosmology, such as the Copenhagen interpretation of the 'wave function collapse', 'entanglement' and the anthropic principle, include interactions that fit the definition of basic operations of magic: magical contagion, 'participation' and the direct 'Self over matter' action. Further, the analysis shows that there are historical and psychological links between magical and scientific types of thinking. Magical thinking operates through symbols and supplies a thinker with creative combinations of ideas, whereas scientific thinking operates through concepts and selects from the above combinations those that fit the observed reality. The criteria for this selection are experimental results and the correspondence of creative combinations with the general context of available knowledge. When modern physics and cosmology entered the domains where experiments are impossible, such as the 'theory of everything' and the origin of the universe, selection of acceptable combinations from the flawed combinations supplied by magical thinking becomes less rigorous. As a result, 'ruptures' appear in the border between magical and scientific thinking, through which magical phenomena trickle into physical theories of the functioning and origins of the physical universe.

Chapter 5 analyses the paradox of the apparently impractical investment of vast amounts of money into problems such as the origin and distant future of the universe and the search for galaxies that are millions of light years away from our planet. Indeed, it appears that, given the existing problems with local wars, refugee crises, species extinction, endemics and environmental imbalance, the billions of dollars invested in the study of the distant past and future of the universe could be better spent. This chapter argues that the cause of this paradox is hidden in the implicit belief of modern humans in the immortality of the human soul and of humankind. Recent studies in psychology have shown that most educated individuals who consciously deny their belief in magic or in God still harbour the belief in the supernatural in their subconsciousness. In the XXth century, studies in quantum physics and cosmology confirmed the ancient hypothesis of the inseparable link between the human mind and the universe. The results of these studies, combined with the subconscious belief of modern people in the supernatural, create the hope that the mind of humankind, and possibly the mind (or soul) of a human individual as well, can be immortal. This implicit hope for immortality makes the study of distant galaxies and the remote future of the universe meaningful.

In Chapter 6, the origins of modern religions are discussed in the light of the beliefs of modern people in the supernatural. Historically, modern world religions (such as Judaism and Christianity) branched off from ancient magic and maintain a strong link with magical reality. Psychologically, this link is supported by the fact that most educated adults still hold a belief in the supernatural. However, for many people today, critical thinking based on the achievements of modern science ousted the beliefs in magic and in God deep into people's subconscious. As a result, there emerged an 'existential vacuum' in the minds of modern secularised individuals and the search for the meaning of life began. Various approaches to this search are discussed, such as 'rational mysticism', as well as attempts to approach magical reality through scientific methods. These attempts include the search for the brain localisation of mystical and religious experiences and experimental studies of paranormal phenomena. The basic need of humans for a meaning to life explains how beliefs in God and in the supernatural survive in the modern, rational world.

Part III of the book (Chapters 7–9) is about applications of magical thinking in various domains of modern life, such as politics, economics, education and medicine. Chapter 7 raises the issue of how a ruling authority can employ people's implicit magical beliefs in order to manipulate the mass consciousness. The structure and functioning of a special cultural–psychological mechanism – 'belief in magic-based social compliance' (BMSC) – are analysed via three contrasting modern cultures: Russia, Mexico and Great Britain. The historical origins of BMSC in various cultures are discussed and the modern condition of this mechanism is assessed. Factors are discussed that could influence BMSC in Russia, with the aim of partial liberation of the Russian people from the grip of unintentional compliance with the suggestive power of authorities.

Chapter 8 examines the role of magical thinking in children's cognitive development. Recent studies have shown that watching a movie with magical phenomena can stimulate children's cognitive functions, such as divergent creativity and visual analysis, to a significantly greater extent than watching a similar movie without magical phenomena. A psychological mechanism of this effect is rooted in the systemic nature of psychological functions. The stimulating effect of watching the magical phenomena creates a psychological ground for developing 'alternative handbooks' on various subjects (e.g., physics, biology and psychology), in which not the laws of science but the laws of magic would hold sway. Handbooks like these would use magical phenomena as 'thought experiments' over reality, which could provide a way of teaching science complementary to that via direct instruction.

The concluding Chapter 9 contains an overview of the ideas discussed in the book and a brief analysis of magical phenomena that were not analysed in the previous chapters. These phenomena permeate politics, economics, medicine, moral behaviour, human relations and neuroscience. Studies have shown that both children and adults in modern industrial countries, explicitly or implicitly, believe in the supernatural. This hidden belief brings about specific effects in various domains of life, from economics to being in love with another person. The wide spread of interactive electronic displays over recent decades has made access to the imagined magical world incomparably easier than before. This boosts the effects of people's implicit beliefs in the supernatural on their everyday behaviour.

Finally, the epilogue presents a thought experiment in which a world free of magical reality is described. This thought experiment brings us to the conclusion that, in this conjured world, many social ills of the real world, such as religious wars, witchcraft, the narcotics trade and suicidal terrorism, would disappear. But this would be a very dull and uninspiring world in which few people would like to spend their lives.

Notes

* Some philosophers claim that Self is an illusion or a byproduct reducible to more basic constructs, such as mechanisms of information processing in the brain[1,2,3]. From the perspective taken in this book, such theories are misleading, because in order to sensibly talk about any cognitive construct, a person first has to have the feeling of Self.

1 Dennett, D. (1991)
2 Dennett, D.C. & Kinsbourne, M. (1992)
3 Carruthers, P. (2011)
4 Subbotsky, E. (2000a, 2000b)
5 https://en.wikipedia.org/wiki/Complementarity_(physics)
6 https://en.wikipedia.org/wiki/Initial_singularity
7 https://en.wikipedia.org/wiki/Quantum_entanglement
8 Chapters 1, 2, 3, 4, 5, 6 and 9 have been adapted from papers published online in *SENTENTIA. European Journal of Humanities and Social Sciences* in 2016 and 2017.
9 A brief version of this book was published in Russian by Directmedia in 2015.

PART I

Magic in the mind

1

THE MAGIC CRYSTAL OF RENÉ MAGRITTE

Art as a window onto the supernatural

Problem

It sometimes happens that after a long flight we wake up in the middle of the night in a room unknown to us and for a few moments are unable to grasp where we are. Everything around us seems strange: the vague silhouette of the door, the moonlight coming through the window and the walls and curtains are all in the wrong places. For some time, while our consciousness is hastily restoring the events of the last 24 hours, we are trying to answer the questions "Where am I?" and "How did I get here?" And even when our memory puts the broken ends together and gives us the answer, the feeling of being in a strange place doesn't quite disappear.

The same feeling I experienced in René Magritte's museum in Brussels. Trees growing from the table set in the middle of a desert ('The Oasis'), a train emerging from a mantelpiece ('Time Transfixed'), a medieval castle on a cliff that is floating free in the air ('Castle in the Pyrenees'), a winged man and a lion on a city's embankment ('Homesickness'). A half-man–half-fish, a half-plant–half-bird… A weird world of images that are both familiar and strange. Magritte acknowledged that he was indebted for his artistic style to the influence of the Italian artist Giorgio de Chirico, who founded a movement in art known as 'metaphysical realism'. De Chirico's paintings are particularly strange and disquieting: deserted town squares and strange juxtapositions of enigmatic objects immerse a viewer into a dreamlike world. According to de Chirico, "To become truly immortal, a work of art must escape all human limits: logic and common sense will only interfere. But once these barriers are broken it will enter the regions of childhood vision and dream"[1]. This theoretical statement doesn't explain why the artistic style that escapes the limits of logic and common sense should appeal to a modern viewer who lives in the world defined by science and logic. The fact is, the appeal is there, and the question is why.

Artworks by de Chirico and Magritte always give me the feelings of anxiety and unexplained nostalgia. Ambassadors of an alien world, these paintings immerse one in that world, which in one sense resembles our everyday world, and in another sense is fundamentally different from it. Looking at these paintings, I sometimes experience the phenomenon of déjà vu – the sense that in the past I have already seen what these paintings show. It seems to me that I have been in this alien world, walked these deserted town squares lit by the sun, looked at these strange objects and sculptures. There appeared in me an unstoppable longing to understand the messages of this unusual and disturbing world. Somehow I felt that understanding these messages would help me find answers to the ultimate questions: "What am I?", "How did I get into this earthly world?", "What is my destiny here?", "Where would I go after death?" But however hard I tried to grasp the meaning of these messages, my efforts were in vain. Only the feeling of disappointment and anxiety and unfinished thoughts remained. And a new question arose: "To what in our inner world are these paintings trying to speak?" Clearly, these painting are not addressing our aesthetic feeling, if under aesthetic feeling the enjoyment of human and nature's beauty is understood. Indeed, by the grace of forms or by the richness of colours, neither de Chirico nor Magritte's paintings match the paintings of El Greco or Vincent van Gogh. Nor do they address our logical thinking, since logical analysis of these paintings (e.g., "loafs of bread don't fly in the sky") brings nothing but trivialities. Yet somehow answering the above question seemed important. Eventually, and in conjunction with my own research, it occurred to me that one possible answer to this question could be found in psychological studies on the belief of modern people in the supernatural and in psycho-anthropological studies of Palaeolithic art.

Reality of the supernatural

One of the most striking human psychological abilities is the ability to get habituated to almost everything. Due to this ability, most of us from a certain age start viewing the world around us as something to be taken for granted, and even a little dull. The same buildings around us, the same sky over our heads – sometimes grey and sometimes blue – all this makes the world and our lives repetitive and poor in excitement. It is hard to explain to most people, scientists particularly, that they live next to the magical, miraculous, supernatural; that their own existence is a fact unexplained. Indeed, one has to make just a slight shift in one's point of view and the magic of the everyday world becomes obvious. Take, for instance, this house, this tree or this cat running in the street. Each of these objects consists of gazillions of physical particles, but what exactly keeps all these particles together so that they don't dissipate into the surrounding medium like molecules of salt dissolving in a glass of water? In regard to simple objects like a piece of iron or a crystal of salt, their solid structure is explained by the known physical forces, such as the nuclear forces and gravitation, but the physical forces are too general and non-specific to explain the constant structure of living organisms. Why do the particles of matter that this cat consists of stay in the cat's body and not dissolve in the air? Why do they stay in separate organs and don't just mix together like the grains of wheat in a bag? Clearly, there must be 'forms' – of this cat and this tree – that keep the molecules in the cat and in the tree. But where are these forms? One can't see or touch these forms; one can only view them in the 'mind's eye', by observing the objects that these forms make possible.

The most wonderful is the fact that objects 'last' – they exist not for a fleeting moment and then disappear, but stay for some time, the duration of which is different for different objects. It is this permanence and stability that the invisible forms give to every complex object, which is the greatest mystery. Even in the emptiness of a vacuum there are little disturbances – so-called 'quantum fluctuations'[2] – but they only last for infinitesimally short periods of time, while complex living structures may last for thousands of years. The fact that objects in the world are stable and unique creates a misleading impression of this world's ordinariness. Of course, if a cat turned into a tree in our full view we would call it a miracle, a magical event. But if we just tried to 'compress time backwards' and look at the world through this inverted 'time lens', we would see that transformations of this kind happen to objects all the time. Single-celled organisms turn into multicellular organisms, animals develop into humans. We call this process evolution, and because we live inside this process we take it for granted. It seems to us that such transformations of simple things into complex ones happen 'on their own'. But, according to the second law of thermodynamics, only the opposite process – the transformation of complex objects into simple ones – can happen on its own. In contrast, transformation of simple objects into complex ones requires some external force that makes such transformation possible and protects the complex objects from immediate disintegration. Darwinian evolution of species may seem a 'blind' process, but how cleverly it is designed! Random mutations of genes, the struggle for survival of the fittest, preservation of useful mutations in a population – these are only some of the necessary elements of evolution. Someone did a very good job in order to create the 'blind watchmaker' (the image of evolution coined by the British biologist and science writer, Richard Dawkins)[3]. Perhaps, the whole of evolution is nothing but a complex 'watch' that, once wound by someone, keeps working by itself. It is also possible that some unknown forces work in evolution that help squeeze time and reduce the astronomical numbers of individuals destined to die for the lucky few to survive. If this is the case, then the blind watchmaker isn't so blind after all. My feeling is that science has not yet said its last word on the structure of evolution, particularly on explaining why, despite the purposeless nature of natural selection, evolution resulted in the emergence of animals as complex as humans.

And the same creative synthesis happens in inanimate matter, where simple chemical elements are transformed into complex ones. Yet in the modern industrial world, the awareness of the fact that the world is full of miracles has become a privilege of children, artists and poets. As children, we believed in miracles. As adults, we could occasionally wake up to this awareness when listening to music by Mozart or watching films by Andrei Tarkovsky, but for the rest of the time, this awareness is hidden from us until we face special moments in life, such as a threat of imminent danger or death.

Apart from the 'everyday miracles' mentioned above, which are in peace with science, there are things in the world that can't be explained by science. Thus, everyone has a soul, but what the soul is, how we got it and what will happen to it after our deaths – to these questions science doesn't know the answers. What is consciousness? Philosophers have been searching for the answer to this question for over 2000 years, but there is still no commonly accepted theory. Where did the universe come from? Despite all the efforts of cosmologists and physicists to explain the origins of our universe, we are still as far away from the answer as people were hundreds of years ago[4]. We don't even know what a random event is. Scientists talk of random processes all the time, but what is a random event? By definition, a random event is impossible to predict, but then it must be a miracle – a 'something' that emerged from nothing. And if a random event is an effect of certain causes, then the event is not random and must be predictable. An example of a random event is a dream at night. Psychologists try to study dreams and even influence them, but why we see this particular dream in exactly this form and at this particular moment is impossible to causally explain. And what is creativity? If a creative process could be understood logically, it would turn into an algorithm and immediately stop being a creative process. To logically understand creativity would be as fatal for creativity as stopping a jet in mid-air would be fatal for the jet.

So, a modern person in the Western world lives in a sort of an 'enclosure'. The world inside the enclosure is known and explained by science and the world outside is unknown and unpredictable. For the people of earlier epochs, the size of this enclosure was tiny, and those people well realised the vastness of the world beyond. They tried to speak with the world beyond by praying or chanting magical spells. But for the last four centuries in Europe, and in other cultures that inherited the European style of thinking, the situation changed drastically. The size of the enclosure grew immensely, and for most people the borders between the enclosure and the world outside went out of view. This happened because of the phenomenal success of science and scientific education. Science denies magic as a false belief. To support its argument, science provided powerful proofs: the Industrial Revolution, the increase of people's well-being and the length of an individual's life, modern medicine and education, flying in the air and space, radio, television and the Internet[5]. Magic couldn't stand such proofs, so it retreated into the 'backyard' of consciousness.

It retreated, but it did not vanish. Wizards, astrologers, and palm readers keep offering their services in the media, and there is no shortage of customers. Still, the role of the traditional magic in modern Western societies is relatively insignificant. Traditional magic takes its strength from people's explicit magical beliefs, and when these beliefs faded, the effectiveness of traditional magic weakened as well. Everyday life's magic (love magic, fate reading, astrology) is cognitively too simple in order to impress a sophisticated modern mind, which is armed with knowledge and logical thinking. More successful are practices that grew out of magic – religion and psychotherapy. Like magic, religion and psychotherapy exploit the ability of the human imagination to affect human thinking and behaviour – the so-called 'placebo effect'[6]. But even these 'babies of magic' occupy a relatively modest niche in modern life. Under the burning sun of science, religion is fading. Psychotherapy pretends to be a science, but most of its methods are indistinguishable from magic[7]; besides, it takes a lot of time, costs dearly and its results are unstable. So, where is a niche for magic in the modern world? Such a niche exists, and this niche is *magical thinking*. It is in the context of magical thinking that I would like to ponder de Chirico and Magritte's paintings.

Magical thinking and the belief in magic

Magical thinking is the *kind of thinking that follows the laws of magic and embraces objects and events that do not conform to the laws of logic, physics, biology and psychology*. People going through solid walls, animals speaking human languages, gods reading human minds and feeding on the smoke of animal carcasses being burned, time travel – these are examples of magical creatures and events. Magical thinking encompasses the world in which there are no inanimate objects or processes. In this magical world, every object and event has consciousness or a soul of its own. Every entity in this world can be spoken to – one only needs to know the language. Early Egyptians, Greeks and Romans communicated with natural phenomena using the language of magic. Even today, in some Catholic cultures, believers sometimes pray for rain.

Although most educated adults today see themselves as nonbelievers in magic (and some also in God), many nevertheless like to 'play' with magic. These 'imaginary games' may give us the feeling of power and importance. Plunging into the magical world, with its strangeness and unpredictability, can help us shake off the dullness and boredom of the everyday world. In our dreams we can travel in time, speak with our late relatives and observe magical transformations of people into animals. Many of us enjoy watching films with magical content (e.g., the Harry Potter or Lord of the Rings series), reading books about shamans and wizards, studying the myths of ancient Greece or Egypt, attending exhibitions with pictures of magical and mythical creatures and events or immersing ourselves in mystical oriental teachings. In our dreams, by using a magic spell we can instantly move to another planet or easily win millions of dollars at a roulette table. Of course, the world of magical thinking is not all fun. Sometimes things can go wrong in the magical world, producing ghosts and dangers, and we wake up in fear. But most of the time in the world of magical thinking we feel ourselves to be in control of events, and we like it. Because our games with magic unfold in the imagination, they peacefully coexist with our beliefs in science and are acknowledged in the Western world as legitimate forms of entertainment, art or dreaming. But the hope of 'playing with magic' without believing in the supernatural is an illusion.

Psychological experiments of recent decades have shown that deep in the subconscious most of us believe that the laws of magic have real power. For instance, one of these laws is 'the law of sympathy'; according to this law, there is a supernatural link (i.e., 'sympathy') between a person and the person's image (see Table 4.1 for more on the laws of magic). In many traditional cultures, people believed "that by drawing the figure of a person in sand, ashes, or clay, or by considering any object as his body, and then pricking it with a sharp stick or doing it any other injury, they inflict a corresponding injury on the person represented"[8], pp. 1–2. With the aim to examine whether the 'law of sympathy' works with educated people today, psychologists designed an experiment; in the experiment, they encouraged participants to throw darts into pictures of good or bad characters. As expected, participants were less accurate at throwing darts at pictures of the faces of people they liked. In spite of the participants' clear realisation that hitting a person's photo with a dart could not possibly hurt this person, subconsciously the participants followed the magical law of sympathy and tried not to damage the photo of a liked person[9].

In the world of magic, the human Self can directly affect matter. The ancients believed that if a person wanted a certain event to occur (e.g., that another person falls in love with him or her, or his or her enemy dies) and the person performed certain spells and rituals, then the desired event might really happen. Thus, in ancient Rome, people used to write damnations or love spells on lead tablets and hide these tablets in certain places in the hope of achieving the desired effects – the death or love of another person – with the assistance of supernatural forces. In order to examine whether people today believe in direct 'Self over matter' magic, participants in a psychological experiment (university graduates and staff members) were first interviewed on whether they believed or didn't believe in magic and then asked to imagine that a professional witch was going to put a spell on their future lives. In one case, the spell was intended to make the participants rich and famous, and in another to make them servants to evil forces. In the interview, most participants denied that they believed in magic and acknowledged that the spell would not affect their lives in any way. Yet when they faced the choice between the two spells, they opted for the good spell and not the bad one. Interestingly, the participants rejected the bad spell only in the situation when their own lives were at stake. When asked what they would recommend another person (a scientist and non-believer in magic) to do in this situation, most participants changed their minds and said they would recommend the scientist to accept both spells in order to prove to him or herself that he or she didn't believe in magic (see Figure 1.1). Unexpectedly for the participants, they behaved as if they believed in the direct 'Self over matter' magic[10].

In sum, studies have shown that, in modern industrial cultures, educated adults subconsciously believe in the supernatural. Like a strange subterranean plant with its roots growing upwards, and contrary to our conscious belief in science, our hidden belief in magic creeps into many domains of modern life: medicine, education, communication, politics and economics[11,12]. But the area in which our subconscious belief in the supernatural plays a particularly important role is art.

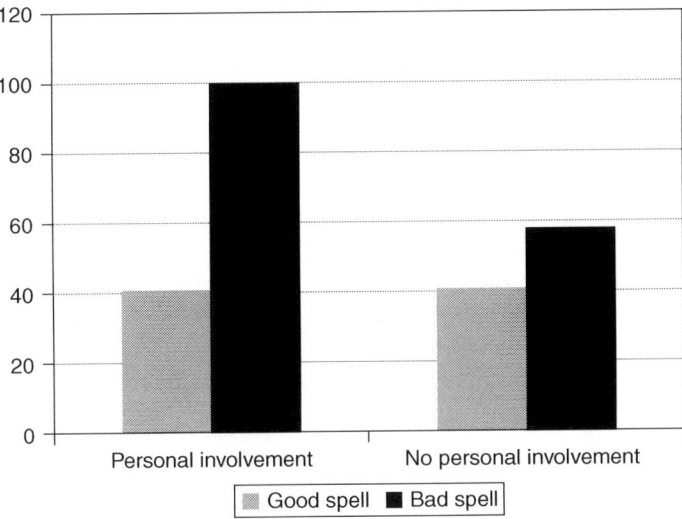

FIGURE 1.1 Percentages of participants who refused to accept the magic spell as a function of condition (personal involvement versus no personal involvement) and type of magic spell (good versus bad)

Source: Subbotsky, 2005

Magical space

Humans broke away from the animal kingdom not when they started making tools and not when they developed language (some animal species, like apes or dolphins, have these abilities, too), but when, with the help of their imagination, they discovered a 'second reality' – the sacred world in which gods and the spirits of dead ancestors dwell – and started to believe in this invisible reality (see Chapters 2 and 6 for more details). This 'magical space' proved to be very capacious: along with being a home for gods and the souls of departed ancestors, magical reality eventually managed to accommodate numbers, letters, schemes, blueprints and nearly everything that is needed for scientific and technological thinking. In order to make this second reality tangible, people developed the first symbols in the form of images, which they carved from wood, stone and bone or painted onto cave walls. It is these early images/symbols that some scientists consider to be the first forms of art[13,14]. In other words, early art, in the form of carved figurines and rock paintings, changed magical reality from something that initially existed only in the imaginations of shamans and wizards into the kind of reality that could be perceived and thus was accessible to everyone. In the course of history, these symbols and images of early art became more abstract and turned into numbers and letters, thus giving birth to mathematics, written languages and science. It seems paradoxical, but in some sense all major branches of modern culture grew out of a common root: they are transformed and diversified forms of magical reality. Like modern continents, which once were one big land mass called Pangaea, *modern art, science, written languages and religions separated from ancient magic millennia ago and have been drifting apart from each other ever since.*

Early burials as forerunners of art

It seems likely that the very first physical representatives of the newly discovered magical world were the objects that our ancestors placed in burial sites. The earliest burial sites are considered to be those of Neanderthals, dating to the Middle Palaeolithic (around 300,000 to 50,000 years ago); however, the absence of any artefacts in these burial sites makes it questionable that the people who were buried there had any idea of life after death. Burial sites that undoubtedly indicate a belief in the afterlife are about 100,000 years old and belong to anatomically modern humans. These burial sites were found in Qafzeh and Skhul caves in Palestine[15]. Along with human remains, the graves contain various objects: deer antlers in the hands of one skeleton, seashells and traces of red ochre on some bones. In later burial sites (no older than 50,000 years), they found primitive decorative objects and hunting tools. It is possible that the first physical artefacts representing a belief in the world of spirits were not carved figurines or cave paintings, but the objects that accompanied the deceased into the world of the afterlife. This suggests that our distant ancestors developed the idea of the parallel magical reality prior to the time when they started to produce the first symbolic artefacts worthy of being called art (such as cave paintings and figurines carved from wood, stone and bone). These ancestors believed that objects that, in the real world, served practical purposes (a stone axe, a bone scraper, a seashell) turned into something supernatural in the sacred space of the burial site, having passed through into the other world together with the spirit of the deceased. Perhaps the ability to view a common physical object from this world (e.g., an axe) as representing the same object that the deceased would use in the world of the afterlife was the initial form of symbolic thinking – the forerunner of later, more genuine symbols: painted images or carved figurines. This capacity for representational thinking was also the first form of sympathetic magical thinking, as it followed the magical law of sympathy: an image (an imagined axe that would serve the spirit of the dead ancestor) equals an object (the real axe that is put in the grave). Even today, in the course of individual development, children begin with using one object as a substitute for another (e.g., a Lego cube as a substitute for a piece of cheese, a wooden stick as a substitute for a horse), and only later become able to draw images (e.g., the piece of cheese or the horse) on paper. So, it is possible that art emerged as a side effect of burial rites.

Yet, the first manifestations of culture commonly acknowledged to be pieces of art were the cave paintings and figurines carved from stone and bone by people of the Upper Palaeolithic. These paintings and figurines contained images of animals and people, and also characters that combined animal and human features.

A drawing as a magical act

Images painted on cave walls by Palaeolithic artists 15,000–30,000 years ago were not pieces of art in the modern sense of the word; they were created with the aim of communication with the supernatural – gods and spirits. According to experts on Palaeolithic art – French anthropologist Leroi-Gourhan [13] and British paleo-anthropologist Steven Mithen [14] – Palaeolithic cave drawings performed magical functions and reflected the anthropomorphic and animistic nature of thinking of prehistoric people. Caves with paintings of animals and people were ritualistic sites in which ancient hunters addressed the spirits of animals they had killed (with the aim of pacifying the spirits) or were hoping to kill (with the purpose of increasing the chances of a successful hunt). Thousands of years later, in ancient Egypt, Greece and Rome, gods were also represented and addressed in the form of painted images or sculptures. Even today, some frescoes and icons are worshipped and believed to have healing powers.

Cave art was the first manifestation of sympathetic magical thinking in people that left tangible traces. This doesn't mean that magical thinking appeared with cave art. In principle, any language requires some kind of representational ability in the creatures that speak the language. For instance, in monkeys, various yells and cries may represent specific dangers (a snake or an approaching jaguar), and every monkey learns that language from an early age. But this language of signals is not based on sympathetic magical thinking. For the language to become magical, the *signals* have to turn into *symbols* – images that represent the invisible spirits of people and animals and addressed these spirits. *Magical thinking, therefore, is rooted in the belief in invisible magical entities* (see Chapter 6 for more on this). In early humans, *spoken language was likely to be the first form of representational magical thinking* (see Chapter 2 for more details), but we know little about this language. Only when people discovered the way of putting spoken words in the form of paintings did sympathetic magical thinking acquire its firm ground – visually presented symbols. The first symbols that developed were pictographs – essentially pictures of objects, like in some Egyptian hieroglyphs. Gradually, these early characters interacted with spoken language and became more abstract. *But even modern languages, as long as they use characters or words that represent something, are transformed forms of sympathetic magical thinking.* The magical essence of a spoken or written word – the belief in invisible spirits – has long evaporated from modern languages, leaving behind only the 'dry skins' of the early symbols. But cave painting show that the first symbols were animated by their spiritual content. This may be a reason why these first images were so perfect.

The phenomenon of Palaeolithic cave paintings' perfection remains an enigma. Images of animals on the walls of Altamira, Lascaux, Shauvet and other caves in Spain and France are unanimously acknowledged as masterpieces of painting that the best artists of the Renaissance and modernity could envy. As the expert on prehistoric art David Whitley put it, with a few exceptions, modern churches, when compared with prehistoric caves, are filled with mediocre art[16]. Palaeolithic masterpieces were created 'at the peak of inspiration', without any sketches or arbitrary lines; the drawings, paintings and carvings were made by single error-less movements of the hand, with no corrections or hesitations. Most surprisingly, cave art emerged spontaneously about 30,000 years ago, without any preparatory stage, and came to a halt equally spontaneously around 10,000 years ago. For the sole purpose of communication with spirits, such grace of form and colour seems unnecessary. Rock paintings of later epochs – and there are thousands of these in various parts of the world – are artistically much more primitive than the magnificent Palaeolithic displays.

So, how can we comprehend this explosion of artistic creativity and mastery in the time of stone tools? Whitley's attempt to explain this phenomenon by assuming that the shamans, who created these images, frequently suffered from bipolar mental disorder (maniac–depressive psychosis), which positively correlates with creativity, is questionable. Evidence for such a correlation is weak; besides, shamans from 10,000 years ago would not have ceased suffering from similar mental disorders, yet the production of artistic masterpieces of the same quality on rock has never happened again. Perhaps prehistoric people better than people of later epochs sensed that contacting the world of spirits could only be done through creating masterpieces of art. But then why don't we see sketches awkwardly and erroneously made and subsequently wiped out, or patched-up lines and colours? Or perhaps the shaman–artists created the images while in an 'altered' state of the mind, and when the artists subsequently recovered their state of vigilant consciousness, they were unable to replicate their own creations. We don't know. All we know is that around 30,000 years ago there appeared painted images that were extremely graceful and performed the function of communication with the magical world. Let us investigate whether this fact can help us answer the question of what makes the 'metaphysical masterpieces' by de Chirico and Magritte so poignantly attractive and mysterious.

The belief in the supernatural and modern art

Stonehenge (a prehistoric stone circle monument in England), the pyramids of Egypt and the medieval gothic cathedrals in Europe are known as masterpieces of architecture. Yet they were created not for enjoyment of their architectural design, but for communication with gods. Modern architectural masterpieces (the skyscrapers of New York, the Eiffel Tower in Paris, the tower blocks of the City of London, the Stalinist skyscrapers of Moscow) by their forms look similar to pyramids and cathedrals, but their functions are purely utilitarian. Why, then, when we are looking at modern skyscrapers, do we still feel aesthetic pleasure? Could it be because these magnificent buildings are reminiscent of ancient magic, of the delight and mystery of communication with the supernatural? Recall that cave paintings and prehistoric figurines, just like pyramids and cathedrals, originally were not the objects of art – they were idols, objects of worship. It was millennia later, in ancient Egypt, when images of gods and spirits (as well as people who were associated with gods – kings and priests) came to be perceived as beautiful and began eliciting aesthetic feelings in viewers. Still later, in ancient Greece and Rome, paintings and sculptures of ordinary people (athletes, fine young women) were included in the family of beautiful objects. This historical link between the divine and the beautiful suggests that aesthetic feeling is a converted form of what originally was the feeling of communication with the mystical world of magical reality, and what we today view as beautiful objects in our distant past were regarded as objects of religious devotion.

This *transformation of the mystical into the aesthetical* happened in music as well. According to Greek mythology, the inventor of music, Orpheus, addressed his songs to gods and spirits. He was able to draw tears of happiness from gods and people and thus to influence them and manipulate them. Being a son of a god, Orpheus knew how to make sounds that the spirits living in trees and stones could hear. Another poet (this time a real person), Homer, begins his poem *Iliad* by addressing the gods: "Sing, O goddess, the anger of Achilles son of Peleus"[17]. Just like the art of painting, music and poetry were originally prayers, ways of communicating with the supernatural. The entrancement that the listeners experienced when communicating with the gods through sounds, by gradually losing its magical context, eventually evolved into a feeling of aesthetic harmony. Music and poetry became art when they ceased being magical incantations.

One more area that grew out of magical communication with the gods was mathematics. In Babylon, ancient Egypt and ancient Greece, numbers were objects of worship; they had a mystical connection with the gods, people's fate and natural and cosmic phenomena[18]. Gradually, the belief in the divine essence of numbers disappeared and people started to view numbers, with the exception of a few (e.g., the numbers 13 or 666), as abstract symbols. But the existence of fundamental constants (such as the speed of light, the fine-structure constant, Newton's gravitational constant, and Planck's constant) in cosmology[19] or the "magical number seven, plus or minus two" in psychology[20] supports Pythagoras' idea that the universe and human consciousness are built on a foundation of numbers[21]. Unlike science, art still holds the memory of numbers being a part of magical reality. It is no coincidence that in some paintings the God-creator is shown holding a compass in his hands. The very process of transformation of objects of religious devotion into the objects of aesthetical admiration happened with the help of a number – the so-called 'divine proportion'[22]. In art, this proportion determines human beauty as well. If you change the 'ideal proportion' of a human body, the beauty of the body vanishes. In a sense, Plato was right: a beautiful person or a beautiful object is beautiful because it is 'touched' by the hand of god, and producing a beautiful poem requires the assistance of divine powers[23].

To summarise, art emerged out of magic, like Aphrodite rose from sea foam. Due to various reasons, in modern urban cultures, the sense of being in touch with magical reality has waned, and the symbolic body of magical reality – paintings, sculptures, architecture, music and poetry – turned into aesthetic reality. Pyramids and cathedrals turned into skyscrapers and tower blocks, images of gods and spirits turned into the paintings of modern artists and magical incantations turned into the songs of poets and the music of composers.

One might ask: "And what about religion? Isn't religious thinking a kind of magical thinking?" From the psychological point of view, it is. Paradoxically, some modern religions, which, like art, evolved from ancient magic, have tried to wipe out all traces of the magical 'umbilical cord' that connected art to magic. In Judaism, Byzantium orthodox Christianity of the eighth to ninth centuries and Islam, painted images of animals and people were banned. Traditionally, this ban is interpreted as the struggle of monotheism with paganism[24]. But in reality it was a fight of consolidated religion with ancient magic, which monotheistic religions viewed as a competitor in the struggle for human minds. Magical reality has always been visually presented to ordinary people through tangible images in paintings and sculpture, starting with Palaeolithic cave paintings and finishing with sculptures of gods in early Egypt and classical Greece and Rome. Paintings and sculptures were symbols that represented the gods and spirits in which people believed. To undermine people's contact with magical reality, it was necessary to destroy symbolic representations of this reality. That is why Islam forbids creating images of living creatures. In Christianity in the Middle Ages, the art of painting was preserved and in the Renaissance period it even flourished. Yet, under the pressure of religion, the link between objects of art and magical reality faded and submerged into the subconscious. It was no coincidence that in most Renaissance paintings and Russian icons, gods, angels and saints were shown in a realistic manner as ordinary people; to reveal the divine nature of these characters, artists had to supplement these images with special features. Angels had wings; holy men and women had a halo (a ring of light) above their heads. There are only a few paintings in which the artistic mastery is so powerful that supernatural characters can be recognised without special features attached to them. This immediate recognition of the divine nature of a painted character we can experience, for example, while looking at Leonardo's 'Mona Lisa'. Like the blood that showed up on the character's hand in *The Picture of Dorian Gray*, in the Mona Lisa, the supernatural seems to trickle from the canvas of the painting. Perhaps the association with the divine is the secret behind the powerful, enigmatic attractiveness of this portrait of an otherwise ordinary young woman.

Another factor that assisted the conversion of the mystical into the aesthetical was modern science, which emerged between the sixteenth and seventeenth centuries. The mechanistic view of the world and a human being, which is characteristic of modern science, undermines the belief in the spirituality of humanity and nature and, as a consequence, weakens the belief in the supernatural[25]. This mechanistic view further promotes the secularisation of modern art. In spite of this, art still maintains its link with the magical world. Science made people's thinking and beliefs more rational, but was unable to change the nature of people's emotions, perceptions and communication. Thus, science rejects the so-called 'argument from intelligent design', which was put forward by the British theologian William Paley in 1802. If, when crossing a field, I came across a stone, Paley reasons, I would be likely to think that the stone has always been there. If, however, I came across a watch, I would think that a craftsman had made it, because it is unlikely for a mechanism as complex as a watch to have emerged on its own. Whether the universe and human beings were created by God (the creationist point of view) or evolved naturally (the evolutionist point of view) theorists still argue, though in science the view prevails that the universe and animal species evolved as a result of chance mutations and natural selection [3].

In art, the situation is different. In order to create a beautiful portrait or an image of woodland scenery, an artist can't be inspired by the knowledge that this person or this wood evolved from the chaos of random processes. Rather, the artist is inspired by the idea that beauty in humanity and nature proves the existence of the divine designer. God created beautiful woodland scenery, and the artist's task is to recreate this scenery in their picture by preserving the scenery's beauty. In this sense, painters of the realistic trend in art, such as Rubens or Rembrandt, are 'intuitive creationists' and magical thinkers.

The way art affects people is similar to magic as well. Art appeals not to our logical thinking, but to our perception and feelings. People's communication through images and language is not based on physical causality (see Table 0.3 in the Introduction). Physical causality is a transfer of a force from one object to another by means of one of the four known fundamental types of interaction – gravitation, electromagnetism and the strong and weak nuclear forces[26]. In contrast, communicational interactions, though they include physical processes – light and sound – transmit not a force but a meaning. For instance, when we ask a person at a table to pass a saltcellar, we cause an action that cannot be reduced to the physical energy that we spend on speaking the words. Human portraits created by great artists speak to us not by words, but by gaze. The philosopher Oswald Spengler called this effect 'an enigmatic action at a distance'. "An enigmatic action at a distance," he wrote, "permeates… from the world of the piece of art to the world of the viewer. Even in the early Florentine and Rhenish paintings… one can still feel traces of this magic"[27], p. 461, translated from Russian by the author]. In other words, a piece of art is a 'wormhole' that connects one person (e.g., a portrait's character) with another (a viewer), one historical epoch (displayed in the painting) with another (the time of the viewer) and one kind of reality (magical) with another (everyday reality).

Artist as an 'avatar' of magical reality

Every artist, explicitly or implicitly, hopes to be an instrument through which the gods (or, to put it in modern terms, our subconsciousness) speak to us. Since the dawn of time, the gods have spoken to people through their chosen ones – shamans, priests and people of art. Even for those chosen, getting access to the gods isn't simple: it requires entering an 'altered state of consciousness'. A shaman enters the state of a trance, in which he or she produces incomprehensible utterances. By interpreting these utterances, the listeners derive directions and revelations received from the gods. The creative work of a modern artist unfolds in a similar way. In a sense, a piece of art is a prayer, shaped in the form of paintings, sculptures, music and architecture. Because the language of the gods doesn't have to be clear to mortals, the artist's creation doesn't have to conform to ordinary logic. The American abstract artist Jackson Pollock 'spoke with gods' by throwing paint on a canvass; others, like Pablo Picasso or the British figurative artist Francis Bacon, were 'listening to their subconscious' in a more comprehensible manner by creating images that remotely resemble commonly known objects. Yet in modernist art (abstractionism, surrealism, cubism, expressionism), most masterpieces are not fully comprehensible either to the viewers or, for that matter, to the artists themselves. Classical paintings, though partially, are accessible to most people's understanding, but the museums of modern art are 'libraries with coded messages' from the world unknown to us. Visiting a museum of modern art, an ordinary person has a sense that the paintings and installations he or she is looking at contain 'something', but for a more profound understanding, most people would need 'an interpreter'. Perhaps modern art accentuates our intuitive association of art with magic, yet what exactly links modern art with the realm of the supernatural warrants an explanation.

Abstract art as a magical experience

If realistic fine art has its clear links to Palaeolithic cave paintings, then what are the origins of abstract art? What relation do seemingly bizarre patterns of colours and shapes, like those created by the Russian artist Kasimir Malevitch, have to the magical images covering the walls of Stone Age caves? It turns out that a thread can be found that connects even these abstract paintings to the magical universe. To begin with, abstract forms have a very old history. Hieroglyphs of Egypt and China, Sumerian writings on clay tablets, abstract shapes painted by aboriginal peoples of Australia and tattoos by peoples of Pacific islands were invented thousands of years ago. Even more old are prehistoric notches carved in wood, stone and bone.

According to ancient Chinese legend, Emperor Fu Xi saw a dragon coming out of a river, mysterious abstract patterns written on its back. These signs, or trigrams, copied by Fu Xi, represented the fundamental principles of reality; they laid the foundation for the philosophy Bagua, which interpreted the structure of the universe and the role of humans in it[28]. The Russian art theorist Paola Volkova[29] draws a parallel between the abstract art movement called suprematism and the magical ancient trigrams. According to Volkova, the suprematists' forms, such as the famous 'Black Square' by Kasimir Malevich, carry a hidden 'impulse of energy' – a capacity to evoke in a viewer the feeling of tension and the appreciation of the beauty of plain, objectless shapes and colours. The aesthetic of suprematism is widely used in modern architecture (e.g., in shaping tower blocks and skyscrapers), technical design and sport.

Extending the parallel made by Volkova, one can assume that the abstract symbolism of our ancestors, which was incarnated in hieroglyphs and trigrams, originally performed the role of magical incantations. In the course of the transformation of the mystical into the aesthetical, this ancient symbolism evolved into the aesthetic of modern abstract art. The provoking patterns of shapes and colours created by abstract artists today have no explicit association with magic, yet implicitly they carry the capacity to trigger in their viewers emotional feelings, which in our distant ancestors were triggered by hieroglyphs and trigrams. Without challenging our scientific worldview, abstract art creates visually coded magical incantations – the 'energy-generating' psychological tools – which elicit aesthetical emotions that earlier were elicited by communication with the magical world.

The aforementioned expert on Palaeolithic cave art David Whitley pointed out that there is a link between the rock art of the Upper Palaeolithic and abstract paintings by David Hockney and Wassily Kandinsky [16]. Whitley believes that a feature common to Palaeolithic and modern artists is synaesthesia – a neurological phenomenon in which stimulation of one sensory pathway leads to automatic, involuntary experiences in another sensory pathway. For a person with synaesthesia, one kind of sensation, such as the experience of colour, might acquire properties of another kind of sensation, such as the sense of smell, taste or touch (e.g., smoothness, roughness or waviness) unique for each colour. Synaesthesia can occur when a person is in a state of trance – an altered state of consciousness – such as those that shamans enter into during their ritual dances, and modern artists can also bring themselves into such states while working on a sculpture or a painting. According to South African archaeologists Lewis-Williams and Dowson[30], Palaeolithic art was created by shamans in a state of trance. In this state, the brain can generate the so-called 'entoptics' – special visual patterns that are caused by processes happening within the observer's own eye (zigzag lines, dots, grids, spirals). When the state of trance intensifies, these entoptic images can grow into sensible shapes that resemble animals, people or fantastic creatures, which neurologists might interpret as visual hallucinations evoked by the trance[31]. When in this state, a shaman enters the realm of magical reality, in which he or she can meet creatures (e.g., half-man, half-beast) and events (e.g., people who can fly in the air or breathe under water) that are impossible in the real world.

Whatever the neurological explanation for the origin of abstract shapes in the mind, psychologically these shapes were interpreted by shamans as proof that they had reached the magical world of spirits and could then communicate with the gods. If the belief in the supernatural still lurks at the bottom of the mind of educated urban inhabitants, then modern peoples' subconscious emotional reactions to the objects of abstract art may be similar to the reactions that our distant ancestors experienced in the presence of the divine.

Conclusion: magical images in the modern world

So, what mystery is hidden in Magritte's paintings? Why, when looking at these paintings, do we experience a disturbing feeling of being on the door-step to an unknown and enigmatic world? Why is there a feeling of déjà vu? Why do we feel that these paintings might give us answers to the fundamental questions: "How did we get into this earthly world?", "What are we here for?" and "Where will we go when we die?"

Trying to find a clue to this mystery, we turned to studies on the origins of art and on magical thinking in modern people. These studies show that at the very beginning of art lies people's belief that next to the ordinary, everyday world there exists a magical, supernatural world in which gods and the spirits of dead ancestors dwell. In order to visually represent this invisible world and communicate with the spirits, people created special means – paintings, sculpture, architecture and abstract signs and symbols. In the course of history, the belief in the magical world of spirits was replaced by the belief in God and, in many people today, by the belief in science. The feeling of the mystical turned into the feeling of the aesthetical – the feeling of appreciation of beauty or good taste – and the means that made the magical world visible turned into modern art.

However, psychological studies on magical thinking today have shown that the belief in the supernatural in modern, educated individuals has not disappeared, but rather has descended into the subconscious. Although consciously we deny that we believe in magic, subconsciously we can't help feeling that beyond the predictable world of everyday reality there exists a mysterious world of the supernatural. In this invisible world, the laws of magic and not the laws of nature hold sway. We had been in this magical world before coming into this earthly world, and we will return to this world after we die. Our education, scientific knowledge and logical thinking have built defences that protect us from this 'memory of the subconscious'. But there is one niche in the modern world in which these defences are powerless, and this niche is art.

In myths and fairy tales, a person can turn into a beast, fly in the air or breathe underwater. Magical creatures similar to these mythological characters abound in modern art as well. Picasso's Minotaur, the bizarre combinations of animal and human features in the 'biomorphs' by Salvador Dali and Francis Bacon, people flying in the sky in the paintings of Mark Chagall – these are images of a magical world snatched by the artists' inner visions from their subconscious. Artists of the Renaissance, with a few exceptions (Breughel, Bosch), portrayed the world by using realistic images; in contrast, in the twentieth and twenty-first centuries, artistic language became increasingly magical. What challenges in the modern world is this artistic language responding to? Perhaps the language of magic is the artists' response to the 'excess of rationality' that surrounds a person in the modern, industrial world. Or perhaps secularisation created a 'metaphysical void' in the heart of a modern, rational person – the void that modern artists try to fill by returning to the ancient language of communication with the gods, the language of magical symbols and images.

It seems to me that the paintings by Magritte and de Chirico are the quintessence of art's historical memory, which takes us back to the art of the Upper Palaeolithic. Through their mysterious paintings, these artists, like no others in the modern world, call for us to acknowledge that what we see in the everyday world – what we believe in – is only the tip of the iceberg. There, beneath the surface of everyday reality, is a magical, supernatural world, and we are not alien to that world. We were there, and will be there again. And our belief in the supernatural, which is hidden deep in our subconscious, silently responds to this call. It seems to us that in this mysterious, invisible world lie the answers to many of the enigmas of our earthly being, and Magritte's paintings right now, in the next hall of the museum, will deliver the answers to these questions. Yet the paintings only lure us further and deeper into the unknown, away from everyday reality, and the answers keep slipping through our fingers.

Did Magritte know the answers? I doubt this, because anyone who knew the answers would not be interested in looking for them. But perhaps, when Magritte died in his Brussels home at the age of 68, he saw the answers.

Notes

1   www.mattesonart.com/magic-realism-giorgio-de-chirico-.aspx
2   http://ru.wikipedia.org/wiki/Квантовая флуктуация
3   Dawkins, R. (1986)
4   Baggott, J. (2013)
5   Horgan, J. (1997)
6   https://en.wikipedia.org/wiki/Placebo
7   Coriat, I. H. (1923).
8   Frazer, J. (1922)
9   Rozin, P., Millman, L. & Nemeroff, C. (1986)
10   Subbotsky, E. (2005)
11   Subbotsky, E. (2011)
12   Subbotsky, E (2014)
13   Leroi-Gourhan, A. (1968)
14   Mithen, S. (2005)
15   Trinkaus, E. (1993)
16   Whitley, D. S. (2008)
17   http://classics.mit.edu/Homer/iliad.1.i.html
18   http://en.wikipedia.org/wiki/Numerology
19   http://ru.wikipedia.org/wiki/Тонкая_настройка_Вселенной
20   http://ru.wikipedia.org/wiki/Магическое_число_семь_плюс-минус_два
21   http://jwilson.coe.uga.edu/EMAT6680Fa06/Hobgood/Pythagoras.html
22   http://ru.wikipedia.org/wiki/Золотое_сечение
23   http://plato.stanford.edu/entries/plato-aesthetics/
24   de Benua, A. (2013)
25   Sheldrake, R. (2013)
26   https://en.wikipedia.org/wiki/Fundamental_interaction
27   Spengler, O. (1998)
28   http://ru.wikipedia.org/wiki/Фу_Си
29   Volkova, P. (2013)
30   Lewis-Williams, J. D. & Dowson, T. A. (1988)
31   Siegel, R. K. (1977)

2

THE INVISIBLE REALITY

Consciousness as a gaze into the magical world

Problem

Imagine that sitting on a bench in a shady park you obliviously slipped into a day-dream. And here, as if with the aid of a magic wand, all the hard problems you have been trying to solve for years are suddenly cracked. You win millions of euros in a lottery, buying yachts, private jets and apartments in the world capitals, the world learns of your existence, newspapers write about you and people look at you with admiration. Gradually, however, your thoughts are switched back to the urgent tasks of the day: you think of what you have to do at work, remember your hard conversation with the boss the other day, make plans for tomorrow. You recall that you had promised to call your sick friend in hospital and that you need to pop into the food store to get some potatoes. Having shaken off the captivating enchantment of the daydream, you return to the mundane and stressful world of *perceived everyday reality* (PER) and have to attend to its demands. When falling asleep late at night, you go through the events of the day in your mind, this time slowly and relaxedly, until the river of dreams hugs you and carries away in an unpredictable direction. Day follows day, and you somehow fail to notice that half of your lifetime, and perhaps more than half, you spend in the invisible reality of the past and the future. The invisible reality of plans, dreams, fantasies and memories, like a ghost, accompanies every moment of your life.

Having looked at this invisible reality more closely, you might notice that it consists of two domains. One is an imaginary copy of PER. Suppose, having arrived at your office, you discovered that you had left your watch at home. While trying to remember where exactly you might have left the watch, you mentally scan every nook and cranny of your flat: is it in the drawer? On a coffee table? On the pedestal near your bed? Your flat with all its content is stretched out before your mental eye like a blueprint or a map, which represents the real flat in one-to-one correspondence.

Let us call this mental copy of PER the *imaginary everyday reality* (IER). IER also includes intersubjective imaginary phenomena, such as scientific theories, secular collective beliefs, social customs, social and moral laws. Imagined objects and events that have no prototypes in the modern world but could be created or could happen in the future (futuristic prognoses, images of technical inventions that do not contradict the laws of nature but are too complex to be created in the current state of technology) belong to IER as well. Finally, IER includes memories of past objects and events that ceased to exist (e.g., memories of historic events, images of people who have passed away). Another domain of imaginary reality contains objects and events that cannot have prototypes in PER. In your dreams, you may imagine yourself having superhuman skills, such as being able to read other people's minds or fly in the air like a bird. In this department of invisible reality, the laws of physics, biology and psychology are suspended: you can travel back in time, ride flying horses, see mythical creatures (e.g., centaurs, mermaids) and talk to gods. In the Introduction, we called this domain of imaginary reality *imaginary magical reality* (IMR). As follows from the Introduction, magical (supernatural) reality is not exclusively imaginary. There exist a number of supernatural phenomena that can be perceived, directly or indirectly, via effects that these events have on perceived objects. This class of phenomena constitutes the *perceived magical reality* (PMR). The interface between domains and types of reality is shown in Table 2.1.

Both IER and IMR can be made tangible in the form of *symbolic representation*, through pictures, movies and verbal descriptions (see Chapter 1). In every bookshop there is a section of 'fiction', with books that contain fictional but realistic events, and a section of 'fantasy', where the books depict fantastical events. Museums, too, usually have separate sections for realistic painters (such as Manet) and painters that depict fantastical, magical events (such as Dali). A similar division into fantastical and realistic departments exists in cinematography. With the onset of the computer era, one more form of visualisation of invisible reality became possible – virtual reality[1]. An important psychological feature of invisible reality is that it becomes a part of social life only when converted into the artificial reality of signs, symbols and images. When we imagine magical events, we too have to convert them in symbols, words or images.

As mentioned in Chapter 1, psychologically people departed from the animal kingdom not when they acquired the ability to use tools, language and social cooperation. Some species of mammals, birds and even insects can use simple tools, communicate through signals and use complex forms of social behaviour[2,3,4]. It is the ability to live in an invisible reality that separated the human mind from the animal one. Animals can use their imagination for solving practical tasks. In Wolfgang Köhler's experiments, chimpanzees were able to understand that they could reach a banana that was hanging high from the ceiling in the centre of a cage if they moved a box from the corner of the cage to the centre and stood on it[5]. Primates can use 'tactical deception' of other animals[6] and even humans[7] in order to get an advantage in reaching food. However, animals' imagination can't go beyond the situation available 'here and now'. An animal is chained to the

TABLE 2.1 The structure of symbolic consciousness

Domain of reality Type of reality	Everyday	Magical (supernatural)
Imaginary	Rational constructions of physical objects and events (schemes, plans, blueprints)	Effects of sympathetic magic
		Fantastical creatures (e.g., centaurs, mermaids, flying horses)
	Scientific theories, secular collective beliefs	Time travel
	Laws of logic, science and society	Ghosts, invisible gods and spirits
	Laws of morality	Aliens from other worlds
	Fictional objects and events that could be created or happen in the future (futuristic technical devices, predicted future events)	
	Memories of objects and events of the past	
Perceived	Physical objects (animated and inanimate) and processes that are being perceived at the present moment and whose rational constructions are known	Gods that can be seen (the sun, fire, a totemic animal)
		Phenomenal subjective experience
		The effect of an observer on quantum events
		Quantum entanglement
		The Big Bang;
		A living cell
		A random event
		An act of creativity
		Intentional movements
		Effects of thinking and wishing on patterns of electrical activity in the brain
		Parapsychological phenomena (extrasensory perception, psychokinesis)

world of perception; only a mental leap into the future could free the animal from this psychological captivity. But a leap like that requires an imagination powerful enough to break away from the ground of perceptual experience. In this chapter, the hypothesis that *it was this leap into invisible reality that gave rise to consciousness (i.e., to 'knowledge about knowledge')* is discussed.

The following questions arise: when in the course of anthropogenesis did people invent the invisible reality? What cognitive prerequisites made this invention possible? What event triggered this invention? What challenges in the lives of early humans did this invisible reality respond to? How did the invention of magical reality change the human mind? How do people manage to tell everyday reality from magical reality?

The puzzle of the Upper Palaeolithic

According to the theory of evolution, the early ancestor of modern humans (*Hominina*) separated from chimpanzees (*Panina*) in the Miocene (about 6–7 million years ago) as a result of an adaptation to a changed environment[8]. One theory suggests that this adaptation was a transition from living in trees to living on the ground (the savannah hypothesis)[9]; another theory argues that the transition went from living on dry land to a semiaquatic existence in shallow waters (the aquatic ape theory)[10]. Eventually, this early ancestor evolved into the genus *Homo*: *Homo habilis* (2.80 million years ago), *Homo erectus* (1.76 million years ago) and *Homo antecessor* (1.20 million years ago). These human ancestors were bipedal and could use fire and tools, yet essentially they still belonged to the animal kingdom. Anatomically modern humans evolved in Africa around 200,000 years ago and migrated out of Africa some 50,000–100,000 years ago. However, for about 150,000 years, the behaviours and tools of anatomically modern humans, with a few exceptions, did not differ from those of their archaic ancestors. Approximately 30,000–50,000 years ago, in the Upper Palaeolithic period, a sudden advance in cultural development was noted in the form of complex tools (e.g., traps for catching animals), cave art, figurines cut from bone and stone, more structurally complex dwellings and trade[11]. It is at this time, and relatively suddenly, that humans developed symbolic thinking and art. And here the question arises as to what caused this Upper Palaeolithic 'cultural revolution', which brought symbolic consciousness into being.

Discovery of personal death and the idea of an afterlife

The founder of British cultural anthropology Sir Edward Tylor proposed that the belief in the world of spirits originated from subjective experiences of early people, such as dreams and hallucinations[12]. Having taken these subjective images for reality, people created the invisible world of gods and spirits and started to believe in this world, thus making an error of judgement. Tylor's theory is an early case of reductionism – the tendency to reduce complex mental functions (e.g., beliefs) to basic cognitive mechanisms hardwired in the brain by evolution (e.g., dreams and hallucinations). Modern British anthropologist Robin Dunbar suggests that religion and storytelling could only emerge on the basis of spoken language, and of all animal species, only humans had language advanced enough to make religion and storytelling possible[13]. In his view, only when people created the imaginary world next to the real world were they able to look at the real world 'outside of the box' and ask why this world is as it is. Dunbar writes about two types of imaginary worlds: the fictional world of storytelling and the 'parafictional' world of spirits. The parafictional world differed from the fictional world in that people believed that the world of spirits existed not only in their imagination, but also in reality. This interpretation of the origins of religion is not without its merits, but it also runs into a problem, because for early humans to develop the language that allowed the telling of stories about invisible entities, humans already had to have the idea of the imaginary world of spirits.

Indeed, what exactly distinguishes human language from animal language? In their communication, animals use a complex system of auditory signals, but these signals are tied to the world that is available in their perceptual field. A scared antelope gives a cry when it sees or smells an approaching predator; an angry, dominant lion lets loose a growl to remind the rest of the pride who's boss. In contrast, a human word carries a meaning that is not chained to the perceptual field. Early hunter–gatherers could speak about animals they were going to hunt or plant roots they would later dig for, without seeing the animals or the roots. But how could the idea of invisible meaning enter the minds of early humans without them already being able to think of invisible entities of some kind in the first place?

A plausible way to account for the origins of human language is to view such language as a result of animistic beliefs. In the beginning, early humans developed the ideas of a soul or spirit that inhabits living humans and the invisible world of the afterlife. Gradually, humans started to extend the area of objects that had invisible spirits inside, bringing into that area animals, plants and other natural objects. Only on this ground could people start thinking and talking about people and objects that were not there by addressing the invisible souls of these people and objects as if the addressees were in fact present. *This identification of the person's or the object's spirit with the real person or object was the beginning of sympathetic magical thinking.* At the same time, *this was the beginning of meaning-carrying words.* It is likely, therefore, that *the first spoken word that was specifically human was a noun, and that noun meant 'soul' or 'spirit'.* When speaking about absent objects, people were addressing the objects' spirits. As animistic beliefs faded, the spirits were converted into meanings – the cognitive constructs that are no longer associated with the invisible souls or spirits hidden in objects. But even today, most languages bear traces of their animistic origins. For instance, we say, 'the sun is rising' or 'the wind is blowing' as if the sun and the wind are living entities capable of active, goal-directed behaviours. The anthropomorphic nature of language unmistakably indicates that once a spoken word was a magic spell addressed to a sentient, supernatural entity – a spirit or a soul. To summarise, *the belief in the invisible world of spirits, or early religion, must have preceded the emergence of meaning-carrying language.* But if people didn't have language before they acquired the idea of the afterlife in the form of the spirits, how could this idea emerge?

Answering this question, American neurobiologists Newberg, d'Aquili and Rause adopted a more existential approach[14]. They see the origin of the belief in the afterlife to lie in the human need to cope with the fundamental unpredictability of life. Unlike animals, humans were able to become aware of the fact that they were mortal beings, and this awareness caused a deep feeling of frustration. Inventing the idea of the afterlife helped to reduce this frustration. Natural selection favoured the individuals who believed in the afterlife over those who did not. Viewing the fear of death as the cause of inventing religion is not a new idea. Among the famous scholars who shared this view were Roman philosopher Lucretius (95–55 BCE), anthropologist Bronislaw Malinowski, physicist Albert Einstein and many others[15]. British philosopher Bertrand Russell wrote, "It is not rational arguments, but emotions, that cause belief in a future life. The most important of these emotions is fear of death"[16]. Let us have a closer look at when and how people discovered that they are mortal, and why this discovery might have made them generate the idea of the invisible world of the afterlife.

Burials with artefacts (tools and decorations) placed in graves are the first available signs suggesting a belief in an afterlife (see Chapter 1). The earliest burial sites of this type are around 100,000 years old and belong to anatomically modern humans[17]. Though disputed, evidence suggests that Neanderthals, who coexisted with anatomically modern humans for thousands of years, also buried their dead[18]. There are observations that even some animals – chimps, elephants and other mammals – act as if they understand that a dead or dying conspecific is in a special state[19]. But the animals don't project the death of a conspecific onto themselves. It appears that only humans had imaginations powerful enough to be able to realise that what happened to their deceased tribesmen would also happen to them. In order to be able to invent the idea of life after death, people first had to *discover that life would end in death for each of them* and be shocked by this realisation. The most important cognitive ability that made such a discovery possible was the imagination, since unlike the deaths of their tribesmen, the person's own death was always in the invisible reality of the future.

To summarise, the gradual and slow development of certain cognitive skills (thinking and imagination) in early humans created necessary prerequisites for the discovery of personal death, and *the feeling of frustration and fear* that resulted from this discovery gave birth to the idea that death is not the end of a person. Leaving his or her dead body behind, a person lives on as a spirit. This was the idea of the afterlife.

With the idea of the afterlife, for the first time in human history, the initially monolithic mind split into two parts: the visible everyday reality and the invisible reality of the supernatural. This was a time in human evolution when the human mind acquired a totally new feature – *consciousness*. We can now define consciousness as *the ability to live in two realities simultaneously: in PER and in IMR*. The acquisition of consciousness was not a momentary event. It may have taken thousands of years before the idea of the afterlife became permanently established in the minds of early humans. In its early stages, consciousness existed as the belief in the invisible world of spirits and revealed itself in the form of ritualistic burial sites with grave goods and early animal cults. Having invented the idea of an invisible entity – a spirit – people started to populate objects of the perceived world with spirits as well. The sun, a mountain, a river, a tree and an animal became bodies with spirits of their own living inside them. Thus, the birth of consciousness was also the birth of a religion we now call animism and the emergence of meaning-carrying language.

Nevertheless, the slow progress of material culture in the Middle Palaeolithic indicates that early consciousness made little impact on human behaviour. Tens of thousands of years may have passed before this early 'pre-symbolic consciousness' acquired a new language – the language of symbols.

The emergence of symbolic consciousness

Analysis of the origin of art (see Chapter 1) suggests that images of animals and people in Upper Palaeolithic cave paintings opened the doors into the invisible magical world where the spirits of animals and deceased ancestors dwelled[20]. It appears, therefore, that the division of reality into the two domains – PER and IMR – that emerged in the Middle Palaeolithic was elevated to a higher level in the time of the Upper Palaeolithic. *A new type of consciousness emerged – the symbolic consciousness*, which portrayed the invisible world of the afterlife through painted or carved images and sculptures. Unlike spoken words, which exist only for a fleeting moment, a painted image lasts. In a painted image, the invisible spirit that was hiding in the meaning of a spoken word found stable ground. Thus, a symbol was born. In the beginning, the symbol was a pictograph – a picture of the object or an animal whose spirit the symbol represented. Gradually, mixing with the spoken word, symbols became more abstract, turning into the signs and characters of written languages and mathematics.

Symbolic representation made the invisible world accessible not only to shamans, who invented the magical world, but also to ordinary tribesmen. Visual images of spirits and magical incantations enabled the possibility to communicate with gods and spirits in the same way people communicated with each other. But there was more to the first symbols than just the tangible representation of (and the means of communication with) the magical world.

The important feature of pictorial (a drawing on a wall or a figurine cut from bone) or auditory (a spoken word) symbols is that they are polysemantic (i.e., they can represent more than one thing at once). For example, a drawing of a mammoth on a cave wall could represent a spirit of the mammoth, which was or would be killed, but it could also represent a living mammoth. It was thus possible for the people to start using symbols for representing not only the divine world of spirits, but also the things that surrounded them in everyday life: Living animals they hunted, animal skins they manufactured, roots and fruits they gathered. Having initially created symbolic language for the visualisation of the invisible and communication with spirits, people discovered that symbols could also represent mundane things. So, instead of dealing with real things (e.g., searching for a real animal to kill), people now could deal with symbols representing these things (e.g., speaking of the animal they were planning to kill). This ability to operate with substitutes of real things opened up an ocean of possibilities, such as remembering, thinking, making plans and building imaginary scenarios. In other words, initially people discovered symbolic language in order to speak with the invisible reality of spirits, but the side effect of this discovery – representational thinking – fundamentally changed the way people processed PER, by moving operations with material objects from the perceptual into the mental plane.

To summarise, the emergence of symbolic thinking changed *early consciousness*, which thus far consisted of only two domains: PER and IMR. On the one hand, in the everyday world there appeared a *new domain of reality* – IER, which contained symbolic representations of the objects and events of PER. On the other hand, portraying IMR through visually presented symbols made the *already existing world* of the afterlife more complex and diverse. This diversification of IMR culminated in ancient Egypt, with detailed descriptions of the underworld, the kingdom of the dead and all the magical creatures and deities that populated Egyptian mythology. In Judaism, Christianity and Islam, the pantheon of traditional deities of polytheistic religions changed into demons and angels that dwelled in inferno and paradise, sharing these invisible realms with the souls of deceased people. Even in modern times, the content of IMR keeps changing, with many rational adults converting gods into technologically advanced aliens or time travellers from the future.

Though much later, another type of magical reality – PMR – was also diversified. Initially, PMR consisted of the 'gods that can be seen' – the sun, fire, a totemic animal. Today, it is hard to find a culture that worships the sun or fire; however, scientific explorations in the last four centuries discovered phenomena that can be observed (directly or indirectly) yet do not obey the known laws of nature. These phenomena include both psychological (e.g., an act of free will, an act of creativity, parapsychological phenomena) and natural (the Big Bang, quantum entanglement, a living cell) objects and events. Altogether, while the contents of the 'departments of consciousness' change with history, the structure of modern individual consciousness remains essentially the same as it emerged in the Upper Palaeolithic, and consists of two domains (everyday and magical realities) and four sections: PER, IER, PMR and IMR (Table 2.1).

Functions of consciousness

Some anthropologists have noted that apart from communicating with the divine, art also had a utilitarian function of preserving information[21]. This conclusion follows from such artefacts as bone plates covered with parallel lines, which probably designated the quantities of valuable objects (e.g., killed animals or processed animal skins). Eventually, signs and symbols designating mundane objects (tools, animals, clothes) developed into a language devoid of reference to the divine and aimed at representing IER.

Although symbolic consciousness is linked to other psychological functions (e.g., sensation, perception, memory, language, thinking and imagination), it cannot be reduced to these functions; it also is not a sum of these psychological functions. Consciousness has a structure and function of its own. As argued above, the structure of symbolic consciousness includes four related but separate sections of reality: PER, IER, PMR and IMR. The crucial feature of consciousness is that it has a bipolar structure, in which the everyday world (represented by PER and

IER) is juxtaposed with the magical world (PMR and IMR); this juxtaposition of two cardinally different types of reality allowed early modern humans to break away from captivity within PER and look at PER outside of the box, as if 'from the perspective of the gods'. This newly formed ability of reflection – looking at PER from the perspective outside PER – drastically *changed human social behaviour* by opening the door to executive control and critical thinking. It became necessary for people to watch themselves in order not to offend invisible creatures – gods and spirits. The invisible and ever-present eyes of the gods made it possible for rules of morality to enter the human life.

Indeed, before the invention of IMR, people's social behaviour must have been similar to the social behaviour in animal groups, being based on instincts and learning through conditioning. The discovery of IMR changed that. There thus appeared controlling agents – gods and spirits – who superseded the power of tribal leaders and were never asleep. For example, if a person didn't share his or her food with other members of the group, this might offend the spirits, who then might punish the offender. The onset of symbolic consciousness, which happened in the Upper Palaeolithic, also *revolutionised human thinking*, moving operations with physical objects inside the human mind and thus opening the way to science.

In sum, consciousness created the ground for morality, science and philosophy, but it also made human life more complicated by forcing people to maintain the division between everyday and magical realities. Having created magical reality, people were now forced to constantly coordinate their own behaviour with that of the creatures who populated IMR. Let us call the process of maintaining the division (the border) between everyday and magical realities '*the effort of reality distinguishing*' (ERD). For a long time in history, the ERD mechanism was imperfect and people frequently conflated the worldly and the divine. This conflation manifested in superstitions, visions, everyday magic, magical healing, witch-hunting and other psychological and social phenomena. Eventually, the ERD mechanism improved and became automatised. In the everyday life of a modern, mentally healthy person, ERD functions subconsciously and is rarely noticed, like we rarely notice our heartbeat. However, life becomes more troublesome when, under certain conditions, the ERD mechanism starts faltering. Let us consider some of these conditions.

Voices of the gods and schizophrenia

The etching by Francisco Goya 'The Sleep of Reason Produces Monsters', depicts a person immersed in deep sleep, his reason dulled by slumber and bedevilled by monstrous creatures that prowl in the dark[22]. From the perspective presented in this chapter, this picture is a symbol of a disturbed ERD, when the magical world trespasses on the world of everyday reality. So, when does the ERD emerge in the course of the individual development of a modern person?

Psychological research indicates that from the age of four to six years, children become able to distinguish between perceptual objects (e.g., a perceived cup), imagined objects (e.g., an imagined cup) and fantastical objects (e.g., a monster or a ghost)[23], yet are tempted to believe in the real existence of fantastical creatures. Further research has revealed that while in 6- and 9-year-old children the border-line between physical and fictional domains of imaginary reality is blurred, adults develop the view that in the realm of fantastical (but not physical) imagined reality objects are free from physical constraints[24]. These data show that the ERD is not hardwired into the brain and matures gradually with age. The development of consciousness that took millennia in human evolution is squeezed into the time span of years in a modern child. Luckily, modern children don't have to invent the idea of the invisible world of the supernatural: this world is made ready for them in the form of fairy tales, toys with magical abilities, pretend play, books, movies and, today, computer games as well. To a small child, even living adults seem almost like gods. Bouncing from their interactions with the supernatural, children quickly develop symbolic thinking, executively controlled behaviour and, eventually, symbolic consciousness (see Chapters 8 and 9 for more on this process). Still, almost until they reach their teenage years, children sometimes confuse everyday reality with the reality of the supernatural. Only in the age of adolescence do children develop the ERD ability in its mature state – and they do need this ability.

In everyday life, when we are doing or seeing something (e.g., speaking with another person or watching a movie), we automatically assess the actions (words) of the other person or the movie's actors as belonging either to ordinary or to magical reality. If we see a person who is talking loudly to an invisible person, we may be in doubt about this person's normal state of mind. This silent assessment of other people's behaviour 'on normality', which is mostly subconscious, is the ERD in action. But have people always been able to employ the ERD as smoothly and effectively as they do today?

American philosopher Julian Jaynes hypothesised that up to approximately 1000 BCE, people lacked the ability to reflect upon their own thoughts[25]. In certain circumstances, people took their own thoughts to be the voices of the gods and obeyed these voices unconditionally. In other words, ancients experienced auditory hallucinations similar to those of today's schizophrenic patients. In this type of mind, cognitive functions were divided between two parts of the brain: one part was 'speaking' and the other was 'listening and obeying'. For instance, the voices of the gods that the characters of Homer's *Iliad* heard represent not a literary metaphor, but an accurate description of the voices that the people of the time described in the *Iliad* heard. When the ability for self-reflection finally evolved, people stopped hearing voices; however, in serious mental conditions, such as schizophrenia, the ability for self-reflection gets blocked and patients begin to hear voices, which they sometimes still attribute to gods[26].

From Jaynes' hypothesis, it follows that before 1000 BCE, belief in a divine reality was already there, but people were unable to keep the visible and invisible realities apart and the ERD mechanism was not yet formed. This hypothesis remains controversial and was criticised for insufficiency of evidence, both historical and neurological[27]. From the perspective presented in this chapter, historically the ERD mechanism must have appeared in people much earlier than Jaynes suggests; this mechanism probably evolved in the Middle Palaeolithic as a result of the belief in an afterlife. The reason for this is that without the ERD, early people would have constantly confused the mundane with the divine, thus making it hard for them to function effectively in everyday life, both socially (e.g., communication during hunting or war) and biologically (e.g., coping with all the chores of daily life). Conflation of everyday and magical realities did indeed happen throughout history, but it happened in the form of superstitions rather than hallucinations. For instance, ancient Egyptians believed in household deities like Bes, the protector of women and children. These deities were represented by small idols and often depicted on household items, yet there was still the understanding that these idols *represented* the gods, whereas the gods themselves lived in a realm of their own. Occasionally, people indeed reported seeing images or hearing voices from the realm of the supernatural (e.g., images of pagan gods, mythical creatures, Christ or the Virgin Mary), but those were exceptional cases and not a stable feature of consciousness. The ancients may have been superstitious, but they were not zombies. They must have felt they were under the constant surveillance of the gods, yet they still had the ability to make free choices.

Nevertheless, Jaynes illuminated important features of consciousness: the division of reality into the ordinary and the supernatural, and the difficulties that arise from the necessity of living in both of these realms at once. Today, life in such a split reality most clearly reveals itself in the phenomenon of *magical thinking*.

Magical thinking and the price of consciousness

As argued in Chapter 1, both children and adults are happy to immerse themselves in the world of magical thinking. We enjoy watching films with magical events, reading books about magic and pondering the mystical paintings by Salvador Dali. In dreams we fly in the air, see animals speaking human languages and travel back in time. In the world of magical thinking, like in a room of distorting mirrors, the everyday and the magical, the possible and the impossible are freakishly intertwined. In this world, the laws of physics, biology and psychology are suspended. We are pulled into this world by its strangeness, novelty and tantalising unpredictability. In the world of magical thinking, we rest from the tiring dullness of everyday life. Apart from providing such a break from mundanity, immersion in the magical world performs other useful functions: it stimulates creative thinking, helps us to get rid of frustrating experiences and gives us a feeling of strength and control over our lives[28,29].

Since magical thinking unfolds in the realm of the imagination, our excursions into the world of the supernatural go well together with our belief in science. Our minds maintain the border between the world of everyday reality, in which things obey the laws of science, and the world of magical thinking, where the laws of magic hold sway. This juxtaposition between magical and everyday realities is important for education, since the laws of science become more salient when contrasted with the laws of magical reality[30,31]. Going through the adventures of Alice in Wonderland or admiring the magical feats of Harry Potter and his friends, a child becomes aware that the everyday world is built on different rules and obeys different laws. This awareness helps children better understand and remember the laws of physics and other sciences. It also facilitates the development of executive functioning – children's ability to consciously control their thoughts, attention and actions[32]. But, as mentioned above, the ability to distinguish between the magical and the physical doesn't come naturally. Studies have shown that before the age of ten years, the ERD is unstable and magical reality can easily trespass on the everyday world[33]. Even in an educated adult, the ERD mechanism can fail, which results in *magical thinking turning into magical behaviour*. Let us consider what happens when the ERD is undermined.

Physical science tells us that we cannot affect inanimate matter in a way other than through one of the four known physical forces: gravitation, the strong and weak nuclear forces and electromagnetism. Indeed, we cannot make the sun rise by just thinking hard about it, or cause a car accident to a person whom we dislike by wishing him or her to have an accident. What physical science doesn't take into consideration is the fact that our thoughts and wishes exist not in the physical world, but in the world of other thoughts and wishes. Yes, our thoughts cannot directly affect physical objects, but they can influence our other thoughts and emotions. Psychological studies have shown that if a person had thought about something happening (e.g., that a certain person would have a car accident) and such an event really did happen, then contrary to common logic the person might develop a sense of guilt and responsibility for the accident. This fusion between thoughts and real events can bring about distortions in people's behaviour by converting routine actions into magical rituals. Distortions of this kind are particularly pertinent to people suffering from obsessive–compulsive disorder (OCD) – a mental disorder in which people feel a need to do certain actions that, in reality, have no effect on their lives. For instance, a person with OCD may feel an urgent need to frequently wash his or her hands or check if a door is locked, because not doing these actions might put the person in danger[34]. Consciously, people with OCD understand that there is no causal link between their compulsive actions and real life, yet they find it hard to abstain from ritualistic behaviour. Although OCD affects only about 2.3% of people[35], in certain circumstances most people intuitively follow the laws of sympathetic magic, according to which a magical link exists between two events or objects that are causally unrelated to one another, such as a person and the person's picture. For instance, one study has shown that drinks that had briefly contacted a sterilised, dead cockroach become undesirable and that laundered pieces of clothes previously worn by a disliked person were less desirable than those previously worn by a liked or neutral person. The study also demonstrated that participants rejected acceptable foods shaped into a form that represented a disgusting object[36]. The results of this study suggest that at the level of the psychological mechanisms of disgust and fear of contagion, which is mostly intuitive, the ERD fails and human behaviour begins to follow the laws of sympathetic magic.

One more condition in which the ERD mechanism relaxes its grip on the mind is dreaming. One way to examine this is to study people's emotional reactions towards magic in the state of sleep. In the Western mind-set, magic is associated with dark forces, yet contains a degree of fatal attraction. This ambiguous attitude towards magic is exposed in some masterpieces of fiction, such as Goethe's *Faust* and Thomas Mann's *Doctor Faustus*. Because of this duality in the cultural disposition, people's feelings towards magic are mixed. On the one hand, people are curious about magic and are eager to experiment with it, but on the other hand, they are fearful that involvement with magic may harbour hidden dangers. In the light of day, when people's consciousness is active and the ERD mechanism is at the peak of its power, people do not usually show any fear of magic, which they view as something purely imaginary. But would they be equally bold when they have to deal with magic in the state of dreaming? To examine this, educated adults were offered a magic spell that aimed to help them to see their chosen dreams at night[37]. Some of the participants declined the offer, but the majority accepted it. The results indeed showed a slight increase in the number of chosen target dreams seen by participants who accepted the offer; however, these participants also saw significantly more scary dreams and nightmares than those who had declined the offer (see Figure 2.1). This suggests that despite their conscious disbelief in magic, when dreaming, the participants were anxious that involvement with magic might have a price to pay, and this anxiety resulted in them having bad dreams. Like the experiments described in the previous paragraph, this experiment demonstrated that, in the state of sleep, the ERD is weakened and people start taking magic seriously.

Importantly, under certain conditions, the ERD can falter in people even when they are in full possession of their conscious critical thinking. These conditions

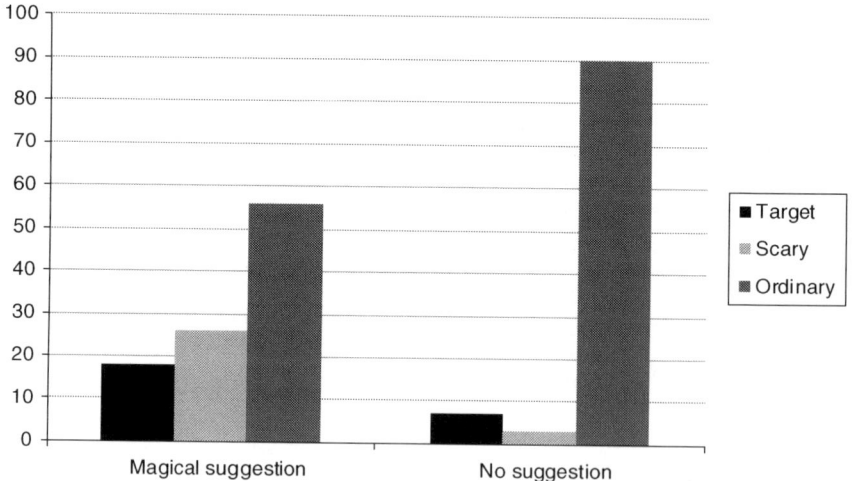

FIGURE 2.1 Percentages of dreams as a function of condition (magical suggestion versus no suggestion) and type of dream (target, scary and ordinary)

Source: Subbotsky, 2009

arise when a certain authority (e.g., a political leader, a priest or a psychology experimenter) claims that magic could indeed work and a person's disbelief in magic is deprived of social support (e.g., when the person has no one around with whom the person might share his or her doubts about the validity of this claim). To examine this possibility, participants (university undergraduates and staff members) were asked to imagine that when they were walking alone in a dark, empty street, a professional witch approached them and offered to put a magic spell on them, which might harm them in the future[38]. Although in an interview the participants denied that the magic spell could change anything in their future lives, they nevertheless acknowledged that they would not allow the witch to proceed with her spell, and explained their decision by admitting that the bad spell might indeed adversely affect their future lives. To their own surprise, the participants discovered that in this situation they behaved as if they consciously believed in magic.

To summarise, the price we pay for consciousness is the necessity of keeping magical and everyday realities apart. Only maintaining this distinguishing effort non-stop, which requires a serious investment of mental energy, guarantees the normal functioning of consciousness. We discussed the conditions under which the ERD starts faltering. But what happens when the ERD fails entirely?

Consciousness disturbed

One of the forms of a failed ERD is *religious extremism*, when the values of the invisible world of the supernatural supersede the values of everyday reality. The psychological and social causes of religious extremism remain unclear[39]. There are reasons to think that religious extremism is a main factor in framing a suicidal terrorist. Psychoanthropological analysis of Palestinians ready to become suicidal terrorists showed that they did not differ from other members of their social environment in educational background, well-being or mental health. The only feature that distinguished radicalised Palestinians from their ordinary compatriots was the intensity of their religious zeal and the belief that their destructive actions were sanctioned by God[40].

The causes of a failed ERD in *mental disorders* have been studied more thoroughly. Male university students who scored higher on the Minnesota Multiphasic Personality Inventory Schizophrenia scale also scored higher on the Magical Ideation scale[41]. Schizophrenic patients showed a higher propensity to resort to magical thinking in reasoning than a control population of similar age[42]. Schizophrenic patients with hallucinations and OCD patients were found to score higher on the scales of superstition and responsibility beliefs in relation to one's own thoughts than the clinical control group and the non-clinical group[43]. These data suggest that a schizophrenic mind is vulnerable to intrusions of magical reality into the reality of everyday life. Moreover, the phenomenology of hallucinatory states in schizophrenic patients shows that the content of magical reality that trespasses into everyday reality is culturally and historically conditioned.

Indeed, Russian psychiatrist Victor Kandinsky (1849–1889) described the hallucinations of schizophrenic patients that he and his colleagues observed; these descriptions show that many of the hallucinations were influenced by the patients' religious beliefs. For example, one of the patients reported seeing a demon who spread his black wings over the whole St Petersburg; another patient experienced a vision that he was looking down at the "abyss of hell" and saw devils going in and out of it; still another patient hallucinated that she turned into an angel, grew wings and "flew a long distance"[44], p. 17. Sometimes, visions of a paradise and an inferno included images of mythical creatures. One patient reported that he was in a paradise that looked like a nicely decorated room in which strange animals resembling a mix of dolphins and small dogs were jumping. Finally, often patients reported having a telepathic communication with God or the devil. But is the content of schizophrenic patients' hallucinations today different from that of patients who lived over a 100 years ago? If it is, then one could expect to hear reports about UFOs and aliens instead of angels and demons. The memoires of Arnhild Lauveng, born in 1972 in Norway, support this expectation[45]. When she was 17 years old, Lauveng was hospitalised with a diagnosis of schizophrenia. Later, she recovered and became a psychologist. She recalls that in her hallucinatory state, she saw various animals (wolves, crocodiles, rats) that sometimes were of unusual colour and size; however, the animals behaved according to their nature. For example, the wolves gnawed on her legs to the bone and the rats would bite, but none of those creatures spoke human languages or could fly in the air. She also heard voices and saw images of people. For instance, a slim woman wearing a dress that was blue and red at the same time represented solitude. The woman was silent and looked like one of her former teachers. Altogether, Lauveng's hallucinatory world looked more like IER than magical reality. However, in her hallucinatory world, there were some details that violated the laws of everyday reality: distortions of space (e.g., a pavement border might seem to be a few metres in depth instead of a few centimetres), unexpected appearances of creatures in places where they were not supposed to be (e.g., wolves in a hospital ward) or incompatible colourings of objects (e.g., a dress that was blue and red at once, orange crocodiles). Other patients in Lauveng's ward saw aliens, Martians or spies.

Lauveng suggests that hallucinations are not arbitrary images, but reflect a person's life experience and symbolically represent some unresolved problems. For example, in her past, Lauveng associated rats with the 'rat race' – the image of 'winning for the sake of winning' rather than for achieving some sensible goal. Lauveng argues that 'races' of this kind are imposed on a person by society, and the rats in her hallucinatory world represented her disappointment with the fact that much of her life was given to this kind of meaningless and forced activity. Interpreting hallucinations as symbols unequivocally links them to magical reality, for the language of symbols is the language of magical thinking (see Chapters 1 and 4 for more on this). More importantly, the content of Lauveng's hallucinatory world clearly shows its link to the cultural–historical background. When Lauveng fell ill, she was a Scandinavian teenager living in a technologically advanced and not very religious country at the end of the twentieth century. It is not surprising that, unlike the hallucinations of schizophrenic patients described by Kandinsky, the hallucinations of Lauveng and other patients in her ward involved images taken from the cultural context of their time (e.g., animals, aliens, Martians and spies) rather than images inspired by religious faith.

These examples demonstrate that an ERD failure in schizophrenia allows magical reality to trespass on everyday reality and be perceived by patients as voices and images. The staff of this 'invasion army' is conditioned by cultural context and the patient's life experience. It is important to note here that a person in a normal state of mind can also experience a strong pressure from images of magical reality (e.g., mirages, visual illusions, stage magic, experiences during film watching and book reading). Yet in a normal state of mind, these images are perceived as illusions and are kept under control. For instance, Victor Kandinsky described 'pseudo-hallucinations' – experiences that feel real but are nevertheless perceived as not real. Pseudo-hallucinations can be experienced by both patients with psychiatric conditions and people in a normal state of mind. Kandinsky argues that 'proper hallucinations' trick perception and consciousness at once, thus making a person view his or her fantasies as if they really did exist. By contrast, pseudo-hallucinations trick perception alone, whereas consciousness keeps functioning normally and qualifies pseudo-hallucinations as fantasies.

Altogether, in both mentally healthy (e.g., religious extremists) and mentally disturbed (e.g., schizophrenic patients) people, the failure of the ERD means trouble, both for themselves and for their social environment. Still, as French philosopher Michel Foucault noted, if in the modern world people who confuse the real with the supernatural are not highly praised, their position is nowhere near as bad as it was a few hundred years ago[46].

A witch as an intruder from an alternative reality

In the Middle Ages, people with schizophrenia were viewed as possessed. With the Holy Inquisition (twelfth to early nineteenth centuries), the ancient hostility of Christianity towards magic became an official institution. Not all those accused of witchcraft by the Holy Inquisition were mentally ill, but all people with schizophrenia were under suspicion of their involvement with black magic[47]. In the Inquisition's view, by treating human illnesses, witches challenged the authority of the Catholic Church, for they fought against the illness, which represented a punishment imposed on people by God for the people's sins[48]. American psychiatrist Thomas Szasz pointed out that magical medicine violated the Church's monopoly on deciding when a person should live and die; by helping the poor and weak, a white witch undermined the traditional hierarchy of medieval society: the authority of a priest over an ordinary parishioner, of a lord over a peasant, of a man over a woman[49]. The episode of confrontation between Simon Magus and Saint Peter described in the apocryphal Acts of the Apostles[50] got a real-life counterpart in the confrontation between witches and the Holy Inquisition.

Yet, despite centuries of persecution, witchcraft in medieval Europe was not eradicated. The reason for this is that for a long time medical doctors were a privilege of rich and famous, while ordinary people were left with no choice but to rely on a sorceress. Apart from magical rites, white witches possessed valuable knowledge of the curing properties of herbs, which may have laid the foundation for scientific medicine. While some of the methods of magical medicine indeed led to science, others relied on the law of sympathetic magic and tried to cure a patient by the transfer of an illness from the patient to another person or animal. Today, the 'magical component' of medieval magical medicine has been condensed into modern homeopathy, but the traditional magical methods of healing are still used by modern practicing sorcerers[51]. Interestingly, in spite of the impressive successes of scientific medicine, psychiatry still bears some features of the medieval attitude towards a patient.

Indeed, in the view of modern psychiatry, mental disorders result from the malfunctioning of certain mechanisms of the brain's chemistry, rather than from possession by evil forces. Nevertheless, as Thomas Szasz noted, the modern clinical assessment of psychiatric patients has something in common with the 'ordeal by water'. The ordeal was used in seventeenth-century England for testing a person's connection with the devil: a person accused of witchcraft was thrown into water with his or her hands tied. If the accused person sank, he or she was considered innocent, while floating indicated witchcraft. The advocates of this ordeal argued that water was so pure an element that it repelled the guilty. In any case, the accused had no chance of staying alive. Similarly, in modern times, a psychiatric assessment of a person suspected of schizophrenia is frequently based on the bias that the subject is ill, which practically guarantees that the symptoms of the illness will be found.

A Google search for 'schizophrenia and witchcraft' brings over 300,000 results. To some extent, in the layman's view, schizophrenics are today still associated with unwelcome guests from the realm of the supernatural. The disturbance of the border between the everyday and supernatural realities still evokes an irrational fear in most people.

Altogether, having emerged in prehistoric times as a person's ability to live in two realities at once, consciousness fundamentally changed human psychology by making human behaviour executively controlled and giving rise to the concepts of morality, freedom of action and personal responsibility. By looking in the distorted mirror of the supernatural, people created art, symbolic language and, ultimately, modern religion and science. But the price for consciousness was high: mental disorders, witchcraft, witch-hunting, religious wars, suicidal terrorism and religious radicalism.

Conclusion: birth of consciousness and the Big Bang

In the beginning of this chapter, we asked the following questions: when in the course of anthropogenesis did people develop the idea of IMR? What caused the emergence of this idea? How did the discovery of IMR change the human mind? How do people manage to distinguish between everyday and magical realities?

Archaeological findings reveal that around 100,000 years ago humans began to bury their tribesmen and put tools, decorations and other artefacts in their graves. This suggests that these people had developed the idea of an afterlife – the invisible reality in which the spirits of the dead lived. Cognitive development provided necessary precursors for the discovery of the inevitability of death; among these precursors, a powerful imagination was a key factor. Of all the animal species, only humans were able to escape from the captivity of the immediate perceptual field and grasp the idea that sometime in the invisible reality of the future every living person was destined to die. The people saw that others with whom they lived, communicated and hunted together would for some reason suddenly become breathless bodies and realised that the same would happen to all of them. This realisation caused an existential shock. By refusing to accept the fact of death, people assumed that the deceased lived on, but left his or her body and passed into another realm with that part of him or her that they called a spirit, and today we call a soul.

Starting from there, it was easy to conclude that gods and spirits possessed unusual properties: they were immortal, invisible, could read people's minds and fed on the smoke of burned sacrificial animals. Having made these conclusions, people for the first time had a chance to look at themselves and their everyday world from another perspective and realised that their own world had very different properties. The ability to see the everyday reality from another perspective, as if looking at it through the 'eyes of the gods', is what today we call consciousness. This ability of reflection gave rise to new forms of behaviour: executively controlled actions and morality. But most importantly, the invisible reality presented a new challenge to humans: in order to deal with this new reality, people had to convert it into something tangible that they could operate. This new challenge resulted in people inventing symbols – tangible images, objects and actions that represented invisible images, objects and actions.

For a long time, people handled the invisible reality by decorating the bodies of the deceased, placing artefacts in their graves and telling stories about the spirits of deceased ancestors. At last, around 30,000 years ago, the people developed a way to 'see the invisible' by drawing images on cave walls and crafting figurines out of wood, bone and stone. There emerged a symbolic way of representing the invisible and communicating with the invisible. Eventually, symbols started to be used for representing mundane objects, such as for recording the number of killed animals or processed animal skins. Both magical reality and imagined everyday reality found their artificial embodiment in the language of signs and symbols. In around 3000 BCE, Egyptians developed written language[52], and in around 2000 BCE in Babylon and Egypt, mathematics emerged[53]. Eventually, symbolic consciousness gave birth to modern logical and scientific thinking.

But consciousness, having pushed humans beyond the boundaries of the animal world, also brought with it psychological problems. The most significant of these problems was the necessity to maintain the border between the ordinary and magical realities. To cope with this necessity, a special psychological mechanism – the ERD – evolved. It took millennia for this energy-consuming mechanism to reach the level of perfection that we enjoy today. In ancient times, and even in the Middle Ages, this mechanism frequently failed, and intruders from the realm of the divine disturbed people's everyday lives. People saw gods, mythical creatures, Christ and the Virgin Mary. Gradually, the work of the ERD stabilised and became automatised by descending into the subconscious. But even in a modern person, the ERD can fail in situations of stress, illness or danger, and when it fails, it opens the door to monsters.

In 1948, the American physicist of Russian extraction George Gamow and his colleagues predicted that the Big Bang must have left in the universe some relict background microwave radiation, which was indeed discovered in 1965[54]. The emergence of consciousness in the Middle Palaeolithic also left a 'relict background' in the minds of modern rational people – a belief in the supernatural. Psychological studies of the last decades did indeed discover this implicit belief. Ousted into the realm of the subconscious by science and religion, this relict belief in the world of the supernatural feeds magical thinking in modern people. By looking into the abyss of the supernatural, people have to, again and again, at every moment of life, generate the subconscious effort of distinguishing between the two realities and thus maintain the life of consciousness.

Notes

1 https://en.wikipedia.org/wiki/Virtual_reality
2 Rumbaugh, D. M., Savage-Rumbaugh, E. S. & Sevcik, R. A. (1994)
3 https://en.wikipedia.org/wiki/Animal_consciousness#Language
4 www.mnn.com/earth-matters/wilderness-resources/photos/15-remarkable-animals-that-use-tools/ants-and-wasps
5 https://en.wikipedia.org/wiki/Wolfgang_Köhler
6 Whiten, A. & Byrne, R. W. (1988)
7 Hare, B., Call, J. & Tomasello, M. (2006)
8 https://en.wikipedia.org/wiki/Human_evolution
9 Leakey, R. (1994)
10 Morgan, E (1982)
11 https://en.wikipedia.org/wiki/Upper_Paleolithic
12 Tylor, E. (1871)
13 Dunbar, R. (2014)
14 Newberg, A., d'Aquili, E. & Rause, V. (2002)
15 www.humanreligions.info/fear.html
16 https://afterall.net/quotes/bertrand-russell-on-the-fear-of-death/
17 Trinkaus, E. (1993)
18 https://en.wikipedia.org/wiki/Neanderthal
19 https://whyevolutionistrue.wordpress.com/2012/08/21/what-do-animals-know-of-death
20 Leroi-Gourhan, A. (1968)
21 Eliade, M. (1994)
22 https://en.wikipedia.org/wiki/The_Sleep_of_Reason_Produces_Monsters
23 Harris, P. L., Brown, E., Marriot, C., Whittal, S. & Harmer, S. (1991)

24 Subbotsky, E. (2005)
25 Jaynes, J. (1976)
26 https://en.wikipedia.org/wiki/Bicameralism_(psychology)
27 Smith, D. (2007)
28 Subbotsky, E. (2010b)
29 Nemeroff, C. & Rozin, P. (2000)
30 Cole, M. & Subbotsky, E. (1993)
31 Wertsch, J. V. (1991)
32 Subbotsky, E. (2015)
33 Subbotsky, E. (2004)
34 Shafran, R., Thordarson, M. A., & Rachman, S. (1996)
35 https://en.wikipedia.org/wiki/Obsessive–compulsive_disorder
36 Rozin, P., Millman, L. & Nemeroff, C. (1986)
37 Subbotsky, E. (2009)
38 Subbotsky, E. (2007).
39 http://en.wikipedia.org/wiki/Extremism#cite_note-ab-1
40 Atran, S. (2003)
41 Thalbourne, M. A. (1994)
42 Tissot, R. & Burnand, Y. (1980)
43 Garcia-Montes, J. M., Peres-Alvarez, M., Balbuena, C. S., Garcelan, S. P. & Cangas, A. J. (2006).
44 Kandinsky, V. Kh. (2007)
45 Lauveng, A. (2015)
46 Foucault, M. (2003)
47 Zilboorg, G. & Henry, G. W. (1941)
48 Michelet, J. (1998)
49 Szasz, T. (1971)
50 https://en.wikipedia.org/wiki/Acts_of_Peter
51 Miller, D. (2011)
52 http://en.wikipedia.org/wiki/History_of_the_alphabet
53 http://en.wikipedia.org/wiki/History_of_mathematics
54 https://en.wikipedia.org/wiki/Cosmic_background_radiation

3

THE BARRIER FOR ROBOTS

Subjective experience as a magical phenomenon

Problem

The 1999 film *The Matrix*, written and directed by the Wachowskis, portrays a futuristic world in which robots have subdued the human population by turning people into sources of energy[1]. In order to ensure that human bodies, which lie in capsules clad in a net of wires, work faultlessly as sources of power, the machines created a simulated reality – the Matrix – which the enslaved humans take for the real world. The film could be interpreted as a modernist version of the theory of the Greek philosopher Plato (427–347 BCE), which suggests that the real world is an illusion. According to Plato, a person is chained in a cave and can only see shadows of the 'real things' thrown by a fire on the wall of the cave opposite him or her. These shadows people take for the real world.

And so, a prisoner of the Matrix, the computer hacker Neo, is immersed in the world of simulated subjective reality and takes this reality seriously. Only assistance from the real world makes Neo learn the truth and so see the real world. In the film, the real world looks unappealing: dead cities demolished by machines, a poisoned ecology, an underground city-sanctuary in which the people who have managed to escape from the Matrix found shelter, and food that is devoid of colour, taste and aroma. A person who managed to move from the world of the Matrix into the real world faces a hard dilemma: to remain in this devastated and colourless world or to return to the more attractive and pleasurable world of illusions.

In the movie's plot, artistic fantasy is skilfully woven into the fabric of the real problems of human existence created by advancing technology. Can robots simulate human subjective experience? Is it really the case that a person who is plunged into the world of simulated subjective reality is unable to tell this world from the real world? What is better: to enjoy life in the illusory world or to move into the true world in order to fight for happiness but live a life of hardship and deprivation?

These are the questions that are discussed in this chapter in the context of recent psychological studies on magical thinking in modern people.

What colour is a magnetic field? The physical world and subjective experience

A blue sky sprinkled with patches of white clouds, the quiet splash of clear ocean waves licking the white sand of a tropical island, the tender petals of a tea rose as if cut from yellowish marble, the delicate taste and aroma of a cream cake – this is how the real world at its best appears to our senses. But those of us who have studied physics and chemistry know that all of the aforementioned sensations are nothing but illusions. In the real world, there are no such things as 'green' or 'blue', 'sweet' or 'sour', 'hard' or 'soft' – there are only electromagnetic waves of certain wavelengths that enter our eyes, molecular structures of food that stimulate taste buds in our mouth and electric impulses that material objects create in special receptors at the tips of our fingers. Reflected from magnolia leaves, electromagnetic waves reach the colour receptors of our eyes and are transformed into neuronal impulses, which proceed further into our brain. Having passed through our optic nerves, the impulses are sucked into the unimaginably complex network of neurons of our brain's visual cortex and – lo and behold – we see the colour green!

This transformation of physical processes into subjective phenomena, or qualia, is a miracle indeed that remains unexplained by science. What existed prior to that moment – light reflected from magnolia leaves, electrical impulses in receptors of the eyes and neurons of the brain – all of these physical structures were connected one to the other by a causal chain, which can be traced back step by step. And suddenly this chain is broken and there appears something entirely different – the experience of greenness. Colours, tastes, odours, sensations of heavy and light, hard and soft, big and small, short and long, quick and slow – the whole world of subjective experiences is a magical phenomenon, if magical phenomena we understand as the events that cannot be causally deduced from other physical structures or explained by the laws of nature (see Tables 0.3 and 2.1). Like in *The Matrix*, we are faced with the following dilemma: which of the two realities – the reality of subjective experiences or the reality of physical processes that correlates with these subjective experiences – is the primary reality and which is the secondary, created reality?

But before we go any further, we need to consider the difference between subjective experience and physical reality in more detail. From the Introduction, we know that the outer world can only reach our Self when dressed in subjective reality. But subjective reality is unstable and prone to variations. For instance, when we are close to a building, we see ourselves smaller than the building, but when we move further away, we can cover the whole building with our palm. So, is the building smaller than our body or bigger than our body? We know that in reality the building is bigger than us, but what does that exactly mean, 'in reality'? There is another problem with subjective experience. We know that subjective experiences in different people are not exactly the same. For example, the table in our study looks average size to an adult who sits and writes at the table, but the same table looks huge to a baby child who crawls under it. When you and I look at the same building, we always see it slightly differently, depending on the angle of view, and sometimes these differences can be huge, such as when I am close to the building and you see the same building from a long distance at the same time. How come, then, you and I agree that we are seeing the same object?

In order to overcome these problems, people invented measurement. Measurement works as follows: you and I agree that this particular object, such as a wooden stick of the length of an arm, is a *measure*. Next, we put this stick to the building in a standard way from top to bottom and designate the building's height, say, as 300 sticks. Of course, to be able to do the measurement, we need to know about numbers, and this was discussed in the previous chapter. Having accomplished the measurement, we created the simplest *rational construction* (from the Latin *ratio* – reasoning). From now on, from whatever distance you and I look at the building, we may *see* its height differently, but *know* that it is always the same – 300 sticks. As a result, every object is now presented in our mind in *two different forms*: as a phenomenon and as a rational construction (see Figure 3.1). *Rational constructions, or knowledge, becomes a part of what we call physical (objective) reality.*

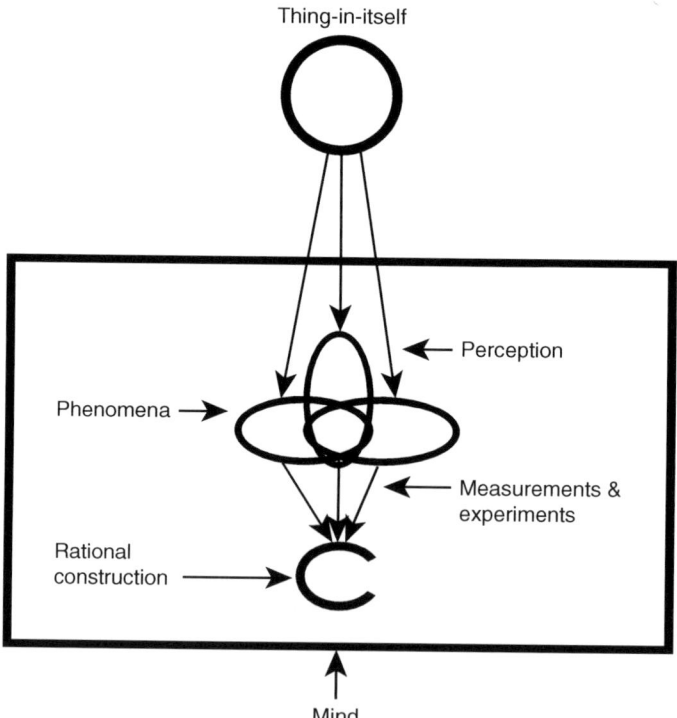

FIGURE 3.1 The relationships between a thing-in–itself, phenomena and rational construction

As a phenomenon, the object varies in size, shape, whether it feels warm or cold, heavy or light and so on depending on circumstances; it also varies between the subjective experiences of different individuals. But as a rational construction, it stays the same for all people and in all circumstances. Science has generated rational constructions for every aspect of objects: size (dimensions), shape (geometrical shapes), structure (molecular composition), weight, temperature, quantities (numbers), space (physical space), time (physical time) and causality (physical causality). Usually, rational constructions exist in the form of numbers, formulae, equations, geometrical shapes and scientific theories. What is important to note here is that the relations between subjective phenomena and rational constructions are not causal, but correlational. The 'real apple' we are seeing and eating is not produced by the apple's rational construction (our knowledge about the apple's shape, weight, size and molecular structure); rather, the real apple (the phenomenon) is caused by the thing-in-itself, via our eyes and brain, and *the rational construction is created by us, through operating with the phenomenon, doing measurements and converting the results in scientific concepts and theories.*

If we are not happy with just seeing the apple but want to find out *how* we see the apple, we look further into the human eye, and see a *retinal projection* – the imprint that the apple as a thing-in-itself makes on the back of our eye. This imprint doesn't look very much like the apple's image that we see: it is much smaller than the apple we see, two-dimensional and upside down (as in Figure 3.2).

If we look still further, things go from bad to worse: the distorted copy of the original apple, imprinted on our retina, is converted into strings of neuronal impulses that already lack the slightest resemblance to the apple we see, and these neuronal

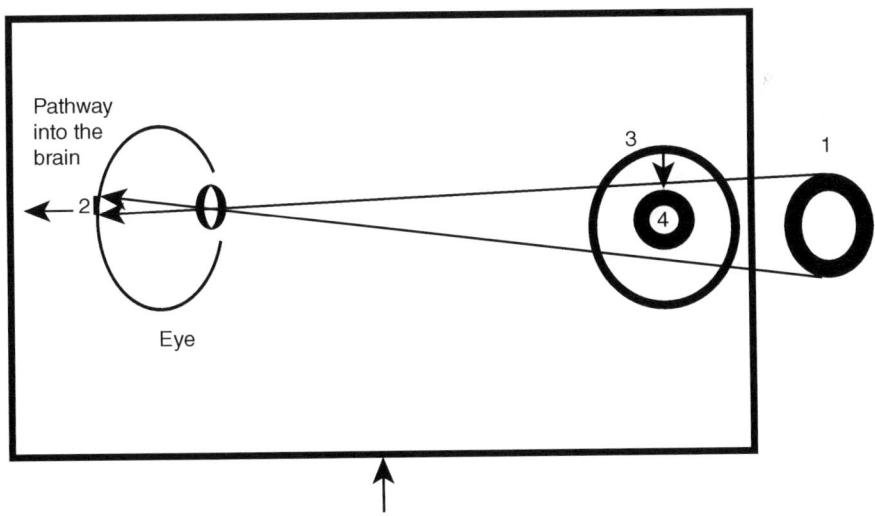

FIGURE 3.2 The relationships between a thing-in-itself (1), retinal projection (2), the phenomenal subjective image that we see (3) and rational construction (4)

impulses disappear into the unimaginably complex labyrinth of our brain's cortex. And then complex conglomerate of neural impulses magically turns into the image of the apple that we see, and we see this image located not inside our brain, but outside over there, where it really is. Even more wonderful is the fact that this *subjective (phenomenal) image has a certain stability, or constancy*; if we increased the distance of the apple from our eyes – say, from one metre to three metres – its retinal projection at the back of the our eyes would diminish three-fold, but the size of the apple we see would stay almost the same. If we looked at a round disc placed in front of our eyes, the disc's projection at the back of our eyes will be a small disc as well; but if we lean the disc 45 degrees, its retinal projection will become an oval, but we will still see a round-shaped disc. This wonderful stability of the subjective image shows that *subjective experience is a medium through which we see things-in-themselves, and this medium has a certain viscosity, or resistance to change, similar to the resistance we experience when we move our hands in water.*

To summarise, the phenomenal subjective experience is a magical medium that *is given to us to perceive the world*, whereas rational constructions *are created by ourselves* in order to make the world of subjective experiences more stable, organised and controllable. Now we can see more clearly that *subjective experience is not produced by brain structures*, because brain structures are part of natural reality (living matter plus knowledge), whereas subjective experience is not a natural entity. However, subjective experience *closely correlates with the work of brain structures*. Still, the question remains open as to which of the two types of reality – subjective experiences or the brain structures – is primary and which is secondary.

It would be logical to assume that the primary reality of the two aforementioned realities is the one that affects the other reality. Indeed, can we change the colour of magnolia leaves or the taste of a cream cake just by thinking hard about changing them? No, we can't. By contrast, we can easily accomplish the above changes by placing the magnolia leaf under a colour filter and by changing the molecular composition of the cream that covers the cake. Not only sensations, but also a lot more complex subjective structures, such as temperament and personality, can be affected by altering material structures. For instance, Huntington's disease causes dementia and a lack of coordination in sufferers, but it can also affect personality, causing anxiety, depression, aggression, egocentrism and worsening addictions, such as gambling, alcoholism and hypersexuality. All of these symptoms result from a mutational change of just a single gene among tens of thousands[2]. Invisible to the naked eye, molecules of drugs, neuromodulators, hormones and viruses, when they enter the brain or body, can change animal and human behaviour. If a capsule of an antidepressant, the neurophysiologist Martha Farah asks, can help us effortlessly overcome problems in everyday life, then aren't personality and temperament nothing but the manifestations of the structure of our bodies? And if this is the case, then is there anything at all in a human being that is not the effect of the structure of our bodies[3]?

Indeed, taken on their own, magnetic and electric fields don't have colour, molecules and their combinations don't have taste, genes have nothing in common with human personality and the chemical substances used in antidepressants have nothing to do with our mood and behaviour. And yet manipulations of these structures (electromagnetic fields, brain structures and genes) can change subjective experiences. It appears that the dilemma "What lies at the beginning: biological structures or subjective experience?" is decisively solved in favour of biological structures. But let's not jump to conclusions.

First of all, it turns out that biological structures are connected to subjective experiences in a very loose way, which allows for a wide range of variations. To start with, it is well known that chemical substances affect different people differently. For instance, the effects of antidepressants depend on the individual structure of the gene 5-HTT, which is responsible for the production and transfer of serotonin[4]. As for the link between our genes and our behaviour, this is even more flexible. Cases like Huntington disease, in which human behaviour can be affected by changes in a single gene, are rare exceptions. Most diseases depend on a combination of dozens or even hundreds of genes. Besides, the way a certain combination of genes affects human behaviour depends on a number of environmental factors. For example, scientists established that hundreds of genes are linked to schizophrenia, yet these genes' roles are mediated by the environment; most often, this unfavourable combination of genes results in schizophrenia in immigrants, especially in those whose culture and appearance differ significantly from those of the indigenous population[5]. There are genes that predispose a person to depression, but whether these genes would or would not activate depends on the person's life experience. If a person's life experience is positive and their number of traumatic events is low, the person is unlikely to develop a depressive psychological condition, even if he or she has a genetic predisposition to depression[6].

Second, psychological experiments have demonstrated that subjective experience could adapt to changes in the mechanisms of the perceptual organs and the brain. For example, wearing special goggles that reverse the retinal image at the back of the eye make participants first see the world 'upside down', but after a few days the normal phenomenal image of the world reinstates itself[7]. When an optical device is worn that inverts the visual field in depth, participants still see certain objects (e.g., a human face) undisturbed[8]. It is also known that people with a damaged brain can recover some of their psychological abilities through training – a phenomenon known as neuroplasticity[9]. If subjective experience were causally determined by brain structures, all these phenomena would not exist.

It turns out that in order to write a computer program for even a few isolated subjective experiences, such as the feeling of depression or the perception of objects, one has to take into account the unique combination of genes and environment. We can barely imagine the vastness of the information about various combinations of genetic and environmental factors that the robots in *The Matrix* would have needed in order to simulate the full-scale subjective experiences of a person, with all of these experiences' colours and sounds, tastes and odours, happiness and unhappiness, hopes and beliefs, love and hatred, sufferings and joy, communication and loneliness. The robots also had to simulate the person's unique personality, temperament, memories of his or her past life, knowledge of languages, the information the person has about his or her country and society and literally all of the bulk of knowledge that a modern educated person holds in his or her mind.

"And so what?" a reader might ask. "The robots of *The Matrix* have almost unlimited capabilities to create the unlimited number of combinations of genes and environmental influences. If the robots were capable of taking the upper hand from their creators – humans – they would certainly be able to simulate full-scale human subjective experience." If the reader's argument is correct, then human subjective experience has nothing to do with magic. In this case, subjective experience is simply a product of a complex combination of genetic and environmental factors, in the same way as the 'ability' of a spaceship to fly in space is a product of a combination of the spaceship's parts. Indeed, none of the spaceship's parts, taken separately, can reach space, but all of the parts put together in the right combination can move astronauts into the orbit. So, let us consider the reader's question more closely.

Brain and thinking: a producer or a receiver?

There are two types of complex systems in a human organism: those that produce and those that receive. For example, the liver produces cholesterol and the thyroid gland produces thyroxin. The liver and thyroid gland are biological systems responsible for the production of hormones. By contrast, an eye receives light and transforms electromagnetic rays into neuronal impulses that proceed into the brain through the optic nerves. The question is: to which of these two types does the brain belong in regard to subjective experiences? Is the brain a producer or a receiver of subjective phenomena?

As far as it concerns sensations and perceptions, the answer is clear: because our perceptual organs (eyes, ears, receptors for hunger and thirst) are teamed up with the brain, they can be viewed as parts of the brain, which are 'brought out' beyond the skull and towards external reality or internal organs. In this regard, the brain is a receiver of sensory stimulation. But how about thinking, personality and emotions? Presently, many researchers believe that thinking is the brain's product. For example, Francis Crick, one of the scientists who discovered the structure of DNA, writes "You, your joys and your sorrows, your memories and your ambitions, your sense of personal identity and free will, are in fact no more than the behaviour of a vast assembly of nerve cells and their associated molecules. As Lewis Carroll's Alice might have phrased: 'You're nothing but a pack of neurons'"[10], p. 3. In this view, the brain produces thinking just like the liver produces cholesterol. But there could be a different view on the relationships between thinking and the brain.

Neurophysiologist David Eagleman offers the following example. Imagine that you are a tribesman living somewhere far away from civilisation (e.g., in a jungle) who has found a sound-transmitting electronic device on the forest floor. Trying various buttons, you suddenly hear human voices. You start to explore how these magical voices could possibly come out of the mysterious box. You open the box, manipulate various wires and see some regularity emerge. You discover that every time you break the contact between a green wire and a base, the voices disappear, but when you restore the contact, the voices return. By manipulating other wires, you learn how to increase or decrease the volume, create and remove noises and so on. In the end, you come to the conclusion that the voices are a product of the enigmatic box. You can even create a theory of how the wires in the box produce the voices, even though you don't know anything about electromagnetism or about the radio station that is located thousands miles away and is presently broadcasting the news or Mozart's symphony into the ether[11]. By analogy, we can consider the relationships between the brain and *complex forms* of subjective experiences, such as *thinking, emotions and free will*.

What if the brain does not produce thinking or emotions, but receives these subjective experiences in the same way it receives stimulation from sensory organs? In reality, the 'symphony of subjective experiences' is being played by an entity that we cannot see, like we cannot see radio waves. Without the radio, we wouldn't hear the 'music of subjective experiences', but the radio is a receiving organ, whereas an 'orchestra' that is invisible to us generates subjective experiences. So, what is the brain: a producer or a receiver? If it is a producer, then subjective experience is a mirage, an illusion, which accompanies the workings of the brain. And what if the brain is a receiver? When a certain part in the receiver fails and the receiver stops functioning, the sound disappears, but not the music. The 'orchestra of subjective experience', which is unique to each individual human being, keeps playing in the magical ether of subjective reality, and can be accessed by you again once the 'radio of the brain' is fixed. Perhaps this orchestra is what we call 'the soul'.

And now suppose that the 'symphony of subjective experience' that the human brain receives is produced by the robots from *The Matrix*. Could the robots create a 'symphony' so complex that a sleeping person would not be able to tell this symphony from real life?

Robots and the magic of subjective experience

Back in the 1950s, Canadian neurosurgeon Wilder Penfield was looking for a way of easing the fits of patients suffering from particularly bad cases of epilepsy. He would access the patient's temporal lobe in the brain and stimulate the cortex with a low-voltage electric current[12]. Penfield registered a new phenomenon, which he named 'double-consciousness'. The patients, who were under local anaesthesia and in full consciousness, reported having two parallel but separate currents of subjective experience. One of the currents was artificially induced by the electric stimulation but seemed to the patient completely authentic, and the other was elicited by the stimuli coming from the environment in the operating theatre. The patients could unmistakably tell the artificially induced subjective experience from the real one. For example, Penfield recorded an occasion in which "a young South African patient lying on the operating table exclaimed, when he realized what was happening, that it was astonishing to him to realize that he was laughing with his cousins on a farm in South Africa, while he was also fully conscious of being in the operating room in Montreal"[13], p. 55. Using the results of his pioneering studies, Penfield developed the cortical 'homunculus map', which showed the cortical localisation of the motor and sensory zones connected to the actions of the limbs and other organs. One might assume that the robots of *The Matrix*, having an immeasurably more perfect 'homunculus map', would indeed be able to create in people the full-scale illusion of real life. But there is a problem.

It wasn't difficult for Penfield to interpret his patients' reports of their subjective experiences, because he knew from his own subjective experience what subjective experience was. But would the robots of *The Matrix* know this? As we know, a person's subjective experience is strictly private; it cannot be directly shared with another person. One can share one's subjective experience only through verbal or nonverbal (i.e., through gestures or facial expressions) reports. But however hard one tries to put into words even the simplest sensation, one is unable to transfer the most important part of it – the authentic quality of this experience, or the 'qualia'. Trying to pass on the authentic subjective experience through words is the same as trying to explain to a person who was born blind what the colour blue is. So, for the robots of *The Matrix* to solve the puzzle of simulating human subjective experience, they would have to have had subjective experience of their own. To acquire subjective experience, the robots would have had to overcome a major barrier – the barrier that separates neural processes (or any physical processes that go on in the robots' hardware) and the 'qualia' of subjective experience. But the robots could not do this without overcoming still another barrier – the barrier that separates living entities from non-living things. In other words, to have subjective experience, the robots have to have become living creatures.

Indeed, today we don't have any proof that an inanimate object, however complex, can feel a need for something, experience an emotion or produce an action of free will. Devices such as thermostats can imitate the actions of a living entity (e.g., acting as if the device 'senses' the surrounding temperature and 'desires' to keep the temperature at a certain level), but in reality such 'behaviour' is programmed by people and only superficially resembles the behaviour of a sentient organism. Only a living entity, starting from a single-celled organism, can behave on the basis of the *inner reflection of its own condition* – the reflection that we call subjective experience.

As for the mechanism of the emergence of life, it is an enigma for science. According to the Oparin–Haldane hypothesis, life on Earth originated in the oxygen-poor primaeval atmosphere that resulted from volcanic activity. Under the impact of lightning and ultraviolet light, organic molecules were synthesised; these molecules later changed into more complex molecules and finally into primitive colloid aggregates called 'coacervates'[14]. In 1953, American scientists Stanley Miller and Harold Urey managed to synthesise in a glass flask simple amino acids, which are parts of protein molecules. They did this by continuously firing electrical sparks into a vapour containing water, methane, ammonia and hydrogen[15]. However, subsequent studies have shown that the primitive atmosphere on Earth contained a significant amount of oxygen, and this would have impeded the synthesis of amino acids. Besides, as American biochemist Jonathan Wells argues, even if scientists were able to create all of the chemical compounds of a living cell, they would not be able to create a cell[16]. Indeed, a modern jet consists of millions of parts, which are put together in a most intelligent way, but this doesn't make the jet an entity that feels a need to fly. The emergence of life on earth, and in the universe for that matter, remains a mystery.

To summarise, subjective experience in its simplest form, such as a feeling of hunger or thirst, can only appear in a living creature. A stone, a river, a planet, a computer – any non-living entity, however complex it might be, is indifferent to its condition. The river doesn't care whether it flows within its usual banks or floods the surrounding fields. A meteor can fly through space for billions of years without the slightest sense of loneliness or boredom. Only a living entity cares about anything. Only a person can ask the question of what caused the flood. Without living creatures, there would be no causality in the universe, and there would be no universe as we know it, for that matter. The ability to have a wish or a drive, to experience joy or pain, to feel the need to explain things is impossible to find in non-living things. This miraculous ability – to be alive and to have subjective experience – is a supernatural ability that is brought into the inanimate universe from elsewhere.

But when and how did subjective reality manage to 'put a claw' into indifferent inanimate matter, to 'subdue' this apathetic combination of atoms? When and how did non-living molecules suddenly start moving not by the order of physical laws, but by the order of an animated entity that exists outside these laws? Perhaps, like subjective experience, life is a magical event and appeared in the universe not because of, but against the laws of nature. If so, then it is understandable that all of scientists' efforts thus far to create or even merely comprehend the origin of life have failed. Perhaps death is a magical event as well. Maybe, when a person dies, he or she does not turn into a mixture of atoms and molecules that dissipate into the vastness of the universe, but passes up onto another level of existence, leaving his or her dead body behind?

We don't know the answers to these questions. Perhaps we'll never know them. But one thing is clear: even the most advanced robots can only imitate living systems, though these imitations can be impressive. It is possible that robots can be programmed to repair themselves and even to create copies of themselves, thus imitating a self-replicating, living cell. But they would only do this because they had been programmed to do so. What robots can't have in principle is subjective experience. But this means that robots will never be able to simulate subjective experience in order to fool a person. Similarly, people would be unable to simulate the subjective experience of an alien creature living in conditions different from those we have on earth.

But let's ask the question: suppose the robots of *The Matrix* were indeed able to simulate the full-scale human subjective experience – why would the robots need to do that? For what purpose would the robots spend so much effort on creating the Matrix when a person's body could be used as a source of energy without the person being in a conscious state of the mind? Indeed, a person who is in a coma-tose state can be kept alive for years, even dozens of years[17]. There can only be one reason for the robots' actions: for their purposes they needed real living people, not just people's animated bodies. And this raises another question: in order to live a full and complete life, why do people need subjective experiences in the first place?

Who moves a finger? Free will as a magical action

With all the diversity of my subjective phenomena (sensations, perceptions, thinking, emotions, actions), the most quintessential phenomenon is my sense that all of these phenomena belong to me. But what am 'I'? Of course, my name, gender, age, family status, nationality, race – all of this is me, but these are only external objective properties. My 'I' also includes my temperament, my personality, my desires, my thoughts and my feelings – but even those properties are not the most essential ones. The most essential property of my 'I' is my feeling of 'authorship' or 'agency' – the feeling that my actions and my thoughts are not imposed on me by someone else, but originate inside myself. The notion 'I am an agent of my actions' means that I am a centre of creation, in which something very real (e.g., emotions, thoughts and actions) emerges from nothing but my Self. People know very well which of their thoughts and actions emerged from their own Self and which they borrowed from someone else. Only those thoughts and actions that emerge from our Self is our real subjective 'I'. But as we know, the emergence of something from nothing is a magical event that violates the principle of physical causality and the law of energy conservation. In philosophy, this magical property of the human mind is sometimes called 'freedom of choice', and sometimes 'free will'.

"But behind every choice," a reader might say, "stands a certain motive or a wish; for instance, when I choose cola instead of mineral water, I give up on my longing for a sweet drink. Certainly, one can't act against his or her drives and wishes." Yes, one can, but we shouldn't conflate an action of *free choice* with an action of *free will*. What is an action of free choice in a nutshell? An action of free choice is the ability to choose from a number of options. Usually, this opportunity turns up when our options are arranged not in a hierarchical order (like a choice between life and death), but laid in a horizontal plain, allowing us enough time to think and discuss the 'pros' and 'cons'. Having considered the options, we finally make a decision: to buy this or another brand of shoes, to order fish or meat in a restaurant or to go for this or another tour. Obviously, the action of free choice only seems to be free, but in reality is determined by our needs. Even so, determination of our actions by our needs is different from physical causal determination. Physical causality works through the known physical forces, such as gravity or electromagnetism. By contrast, determination by needs conforms to magical causation through participation (see Table 0.3 of the Introduction). I submit to my needs because I don't distinguish my needs from myself; my need for food or love is a part of me. Yet my needs are only a part of my conscious 'I' and determine my actions only partially. The other part of my conscious 'I' is my voluntary decision to accept the demands of my needs, with the right to 'veto' these demands. Under normal circumstances of everyday life, the right of veto is rarely used. But when a situation arises in which my private needs come into conflict with my moral consciousness, the action of free choice may grow into the action of free will.

So, what is the action of free will? The action of free will is the ability to choose an option that stands at the bottom of the hierarchically arranged 'survival scale' wired into us by evolution (e.g., to starve to death for a cherished idea or to be honest when there is an opportunity to lie and get rich without undesirable side effects). An action of free will works against one's survival instinct; this action doesn't have a motive except maintaining the awareness of being fair and right. The action of free will breaks the psychological causal chain of 'personal interest–wish–action'. But is the action of free will indeed independent from the brain–body mechanisms, or is it only an illusion?

Searching for an answer, American physiologist Benjamin Libet conducted a series of experiments. In one of these experiments, participants were asked to press a button with a finger or simply to move a finger when they wished. On top of that, the participants were instructed to remember the position of a running dot on a tableau of a device measuring time in milliseconds at the moment "when they only start feeling the wish to make the move." It turned out that the awareness of the wish to make an action and the action itself are separated by a time period of approximately 200 milliseconds. This result wasn't surprising. What was surprising, however, was that an electric potential in the motor cortex of the participants' brains (called the 'readiness potential') appeared 300 milliseconds prior to the awareness of the intention to make the movement. It looked as though the brain 'made a decision' to make the movement prior to the moment when our conscious 'Self' made this decision[18]. If the action of free will is a conscious decision to make the action, then that freedom appears to be an illusion. It looks as though an action of free will is simply a delayed awareness of the decision that has already been made by our brain[19]. But let's look at this experiment more closely.

First of all, we notice that the participants' actions in Libet's experiment were not actions of free will; the participants' actions 'to move a finger' were conditioned by their motive to follow the experimenter's instruction. This motive in its own right was conditioned by the participants' consent to participate in this experiment, which was also conditioned by some other motive (e.g., to get paid, etc.). In other words, Libet studied not the action of free will, but the action of free choice between two options: to move or not to move the finger in this particular moment of time. As we know from our everyday experience, the action of free choice is not a momentary event, but is stretched over time. Usually, when we have a range of options, we hesitate for some time before we make a decision, and when the decision if finally made, the awareness of this also comes gradually. For example, in the case of Libet's participants, they may have hesitated for some time on "Do I want to move my finger 'now' or don't I?" Finally, when the decision of "I certainly want to move my finger now" comes, the participant registers the position of the dot, and after that moves the finger. So, the 'readiness potential' of the participant's brain may have occurred not prior to the participant's conscious decision, but prior to the participant's 'final decision' to proceed with the action. In reality, the readiness potential might have accompanied the participant's period of hesitation, when the participant was aware that they wanted to make the action, but was unsure that they wanted to do it 'now'. The above considerations undermine the conclusion that Libet's experiment testifies to the 'brain's priority' in the action of free choice. To an even lesser extent, Libet's experiment can be used to understand the action of free will. The action of free will occurs in situations of an existential choice, such as a choice of compassion versus egotism, honesty versus dishonesty, moral integrity versus corruption. Studying the action of free will requires a special type of psychological experiment[20,21,22]. These experiments showed that, for both adults and children, making an action of free will is hard but possible.

But suppose that the action of free will, and the whole of consciousness for that matter, is indeed an illusion that reflects, with some delay, decisions made by the brain and the body. Let's investigate if the assumption 'the brain is a decision maker' is free from logical contradiction. Indeed, the question arises of why we have the feeling of agency if this feeling is nothing but a useless reflection of the 'brain's working'. Would it not be more economical for a person to act like a zombie?

One answer to this question is that humans need consciousness for cognition and exploration of the world. Thus, David Eagleman suggests that consciousness is a long-term planner, the CEO of the company, while most of the day-to-day operations are run by those parts of the person's brain to which the person has no conscious access [12]. Indeed, any training, such as learning sport skills or driving skills, starts with a conscious effort that eventually becomes automatised and no longer needing conscious control. By analogy, an action of free choice is only needed in new and unexpected situations. For instance, when you are running down a stony hill, you don't think about where to put your feet; your feet automatically find the right positions. But now you run into a crack in the ground that you can't jump over; immediately you consciousness is switched on and starts looking for a way around the obstacle. In other words, for solving complex cognitive problems we use our ability of free choice, whereas dealing with smaller and simpler problems we pass down to our brain.

The argument that for problem solving consciousness 'comes first' and the brain 'comes second' may sound somewhat unusual, because in the present time the notion that people think 'with the brain' is so common that it has almost become a habit. Yet Greek philosopher Aristotle, who lived in the fourth century BCE, was of the opinion that thinking was the function of the heart and the brain was a machine for cooling the blood[23]. Our knowledge about the brain, nervous system, neuronal chains, etc., which today looks unquestionable and even self-evident, in earlier times was a subject for observation, analysis and conscious decision. In order to generate the idea that subjective experience is an illusion reflecting decisions made by the brain, people had to first use their subjective experience to create theories about the brain. Clearly, the argument that the brain is a decision maker runs into a circle: to discover that the brain is a decision maker, which creates the illusion of subjective experience, we need the support of our subjective experience.

The primacy of subjective experience in relation to concepts of physics and physiology is also obvious from the fact that certain subjective phenomena don't cease to exist after their illusionary status is explained by science. For millennia, people believed that the sun revolved around the earth, until Copernicus explained that this illusion was due to the earth's rotation around its axis. Nevertheless, the illusion is still there: in everyday life, we speak of sunrises and sunsets. Explaining sensations of taste and odour by chemistry and physiology does not change the fact that salt is salty and sugar is sweet. To the luck of specialists on the physiology of colour vision, their theoretical explanations of how colour vision works doesn't deprive them of the pleasure of enjoying the sensations of a blue sky and green grass. By analogy, psychologists and philosophers who think that the sensation of the freedom of action is an illusion don't cease to feel that they themselves, and other people especially, are responsible for their deeds. In other words, subjective experience is causally independent of a rational explanation of the physical mechanisms that correlate with this experience. This means that *subjective experience must have special properties that distinguish it from physical reality*. Let's consider some of these properties.

Properties of subjective experience

The first and foremost property of subjective experience that distinguishes it from physical objects is that *it makes us human*. Any physical part of our body is a means to an end; we need our tongue for speaking, out heart for pumping blood around our body, our legs for walking, etc. This does not apply to our subjective experience. It is not that a person needs subjective experience in order to accomplish something; a person *is* his or her subjective experience. Without a person's subjective experience, there is no the person, there is only what we call 'a body'. But again, a body is a perceived phenomenon and a scientific concept; for both of these manifestations of the body we need subjective experience, either in the form of perception or in the form of thinking. From this, it follows that *subjective experience is a magical phenomenon that cannot be explained by logic and science, but gave birth to logic and science*. Subjective experience is a gift to human beings, and to all living creatures for that matter. A person can accept this gift with gratitude or refuse to accept it. Having accepted the gift, we can develop or spoil it, but we cannot create subjective experience.

A reader might ask, "And what about the fact that, by stimulating parts of human cortex, it is possible to create a stream of consciousness, like the aforementioned Penfield experiments showed? Isn't this artificially created stream of consciousness a piece of subjective experience?" Yes, as argued in the beginning of this chapter, eliciting subjective phenomena is possible, but this would be eliciting, not creating. That is why a person who experiences artificially elicited subjective phenomena can distinguish these phenomena from the parallel real ones, just like we distinguish dreams from reality or a fake from a real thing. But most importantly, artificially elicited subjective phenomena are as inaccessible to rational explanation as are authentic subjective phenomena. We understand how the stimulation of the cortex by an electric current is converted into the person's experience of "laughing with his cousins on a farm in South Africa" no more than we understand how light waves of a certain frequency are converted into the sensation of the colour green. Phenomena such as the feeling of desire, free will, intentionality, voluntary action, creative insight and other manifestations of subjective experience are inexplicable in terms of physical causality; they are magical phenomena of the 'emergence/ vanishing' type. These phenomena correlate with certain processes in the brain and are distorted or vanish when the processes in the brain go wrong, but subjective phenomena cannot be deduced from the brain processes in causal or logical ways.

"Fine," the reader goes on, "but what about *subconscious processes*? Aren't subconscious processes a part of subjective experience that are inaccessible to conscious awareness? And could we not deduce conscious subjective experiences from our subconscious processes?" Indeed, our subconscious memories, thinking and feelings are parts of our subjective experience. Psychologists and philosophers often use subconscious subjective experience in order to 'scientifically' explain phenomena that are not yet fully explained by science. Dreams, hypnotic states, hallucinations, telepathy, telekinesis, Freudian 'complexes', Jungian 'archetypes' and many other unexplained phenomena are relegated to the department of subconsciousness. 'Explaining' unexplained phenomena by placing them into the subconsciousness sounds scientifically plausible and gives these phenomena a legal status in the modern world. But what subconscious subjective experience really is nobody knows. Is subconscious subjective experience a source of conscious subjective experience or is it a 'wormhole' through which magical phenomena of subjective experience filter into our conscious mind? In any case, subconscious subjective experience is as mysterious and irreducible to body or brain functions as is conscious subjective experience. All we currently know about subconscious subjective experience is that it stores information, supports creative magical thinking and makes our body alive. Most of our body functions (e.g., blood circulation or food digestion) are controlled by subconscious processes. When, under general anaesthetic, we don't have conscious subjective experiences, all our body functions rely on subconscious processes. Similarly, a person in a vegetative state can be kept alive for years without conscious subjective experience. When we talk about subjective experiences in lower animals and even in single-celled organisms, we too mean subconscious subjective processes.

To summarise, conscious subjective experience makes one a conscious agent and subconscious subjective experience stores memories, does creative jobs and makes a person's body alive, but neither of the two kinds of subjective experience is the product of the brain or the body. This distinguishes subjective experience from physical or biological phenomena, which always result from certain causally detectable processes. Knowledge of these causal processes allows us to manipulate physical and biological phenomena, such as transplanting a kidney or a heart. But it is impossible to transplant subjective experience.

The second property of subjective experience that distinguishes it from physical reality is that a person's subjective experience *is not accessible for direct observation from the outside*. Indeed, physical objects we first register in perception: this is a stone, this is a neuron and this is a trace of an alpha particle. Having perceived these objects, we compare them one with the other, measure them and assess them using figures. Having done the measurements, we create scientific concepts of physical objects and proceed with establishing causal connections between these concepts via four known physical forces: gravitation, the weak and strong nuclear forces and electromagnetism. By contrast, we cannot see what another person is seeing or thinking, we can only infer the other person's subjective experiences from that person's behaviour. We know that every person has to believe in something (e.g., in God, science or materialism). The person's beliefs reveal themselves through the person's verbal or nonverbal behaviours. By carefully observing and analysing these behaviours, one can study the person's beliefs and then use these beliefs as a 'carrot and stick' to influence the person. This means that it is possible to study subjective experiences by objective methods; it is also possible to influence people's subjective experiences, but not through tampering with their brains. A person's subjective experience can be influenced via manipulation of his or her beliefs, desires or perceptions. Religious leaders, politicians, psychotherapists, specialists in advertising, teachers and artists routinely influence people's subjective experiences without intervening in their brains. Concerning interventions in brain functioning, these interventions affect not subjective experiences as such, but the process of the brain's normal functioning. The work of a neurosurgeon can be compared to the work of a radio engineer who is fixing a radio. Without the normal functioning of its mechanisms, the radio won't play music. But the engineer can't make the fully fixed radio play music if there is no music in the ether for the radio to play.

The third property of subjective experience that makes subjective experience different from objects of physical reality is that, whereas a person's subjective experience is inaccessible for observation from the outside, *it can be accessed by the person 'from inside'*. The fact that we have our subjective experiences inside our minds opens up a unique opportunity to study these experiences in a different way from the way we study physical objects. Indeed, as mentioned above, the first step in studying physical objects is to register them in observation through perception. By contrast, our subjective experience can't be registered in our perception. We see a tree, but not our perception of the tree. We cannot perceive our thoughts, our voluntary decisions or our feelings of love, but we can register these subjective experiences through self-reflection, give them names and interpret them. In the realm of subjective experience, a mental image of an object, a thought about the object and the object's name are connected via *participation*, or, to use a term from quantum physics, they are 'entangled'. For instance, the image, the thought and the name of a rose are not the same, but it is hard to separate one from the others: where there is the image, there is always the thought and the name, and vice versa. Moreover, one kind of thought and image can trigger another kind of thought and image by the same principle of participation. For example, while walking in a park, we suddenly catch the aroma of cherry tree blossom, and this olfactory sensation may trigger thoughts and memories about the events in our childhood and, travelling further along the associative chain, bring us to the most unexpected thoughts and images. The study of subjective experiences 'from inside' is most skilfully conducted not by scientists, but by writers; Marcel Proust's masterpiece *In Search of Lost Time* is one of the most well-known examples.

Finally, one can distinguish subjective experiences from physical objects by the way people exchange their subjective experiences with other people. Physical objects interact via the aforementioned four fundamental physical forces, and their interactions conform to the laws of physical causality and energy conservation. By contrast, *exchange of subjective experiences between two people conforms to the law of participation*. For instance, by saying our thoughts out loud, we address a partner and tune the partner's subjective experience to similar thoughts and images. Suppose a husband, when walking with his wife along the shore of the Red Sea, says, "A beautiful sunset, isn't it, darling?" and the wife answers, "Do you remember our trip to San Diego? There were sunsets like that there as well." Clearly, this exchange of communicative messages cannot be reduced to the exchange of light and sound waves that the speakers produce; the content of the communication follows the magical law of associative participation.

In contrast to associative thinking that conforms to the law of participation, scientific thinking follows the laws of logic, which are similar to causal laws of nature. Logical thinking links subjective experience to physical objects (see Tables 0.1 and 0.2). Logic and mathematics are parts of the external world, designed to create rational constructions and stabilise the diversity and flexibility of subjective experience. Robots are part of physical reality (matter plus knowledge) – a product of logical and mathematical thinking put into a body of complex machinery. This makes it impossible for robots to go beyond the world of physical reality.

Altogether, the differences between subjective experiences and physical reality make it impossible for the robots of *The Matrix* to imitate subjective experiences. Nevertheless, there is a trend in popular culture to ascribe subjective experiences to robots. Let us consider this trend and some of the problems that the anthropomorphisation of robots raises in connection with the topic of this chapter.

Humanisation of robots

Speaking about robots, we sometimes use anthropomorphic expressions, such as 'to teach a robot to do something' or 'the robot can recognise human faces'. Clearly, humanisation of robots goes far beyond the common anthropomorphic nature of human languages. For example, when we say 'the sun is rising' or 'the wind is blowing', we understand the metaphorical meaning of these expressions and do not believe for a second that the sun or the wind are animated entities capable of acting on their own will. However, when we speak about robots, we indeed mean that robots will be capable of setting goals for themselves, if not now then in the future, and consciously achieving those goals.

The phenomenon of the humanisation of robots looks even stranger in light of the fact that nobody doubts the obvious fact that a robot is nothing but a machine programmed by people. Everyone knows that even the most complex modern computer is a tool used for assisting human actions, and in this sense a robot is no different from any other tool, such as a hammer. It is unlikely that someone in the mass culture would say 'teaching a hammer to put nails into a wall'; rather, people would say, 'teaching a person to hammer nails into a wall'. This raises the question: what is specific about robots that distinguishes them from a hammer and makes us apply to them terms that are usually applied to people?

Perhaps we treat robots like humans because some of them look like humans. Many modern robots indeed are designed to look like people: they have a head, hands, legs and a trunk. But let's imagine that somebody created a hammer that looked like a person: would people apply terms of human psychology to this 'humanlike hammer'? Obviously they wouldn't. Besides, not all robots have a human appearance. For example, complex computers that can beat a world chess champion don't look like humans, yet in the mass media they are treated in the same way as humanlike robots. The humanisation of robots has given rise to a popular discussion on the limits of the artificial intelligence (AI) development, and these limits are becoming increasingly vague. In the movies *I, Robot*, *The Terminator* and *The Matrix*, robots frequently outperform humans in their ability to consciously think and act.

Of course, a typical robot today is a lot more complex than a hummer or even an automobile; it is a device full of electronics combined with sophisticated mechanics. Could it not be, then, that it is the complexity of the structure that makes a robot seem similar to people? Again, the answer has to be 'no'. There are complex electro-mechanical systems that are not perceived as humanlike entities, such as modern jets. Flying machines, such as planes and drones, are in a constant process of improvement, yet no one would say that a plane or a drone was 'taught to perform new actions and skills'. And vice versa, some robots with a simple internal structure (e.g., dolls that are used as sexual partners) can be treated as humans and even elicit in people a feeling of attachment.

So, if the features that make us perceive robots as humans are neither the robots' anthropomorphic appearance nor the complexity of their internal structure, then what are those features? The most likely answer is that we view robots as independent subjects because of the functions that robots perform in our lives. Indeed, unlike simple tools and machines that amplify and improve our lower psychological functions (e.g., manual actions and movements in space), robots imitate and improve our higher psychological functions, such as thinking and emotions. It is common knowledge that computers can do calculations a lot faster than humans and can store vast amounts of information. Today, computers have started to imitate even human emotions and communication. Some AI experts suggest that in the not-so-distant future robots will be able to perform the functions of human sexual partners and even become objects of love[24,25,26]. This will happen not only because robots will look increasingly like people, but mainly because they will become able to simulate human emotions and even human individuality and unpredictable behaviour. Even today within the sex industry, dolls are manufactured that can be used as sexual partners for people and elicit in them feelings of being in love[27].

Studies have shown that people are able to develop attachments to robots. In one of these studies, participants were given an opportunity to interact with a cute robot that imitated a baby dinosaur. When the participants were subsequently encouraged to hit the toy, a significant number of participants reported feeling as if they had been asked to hit a living creature[28]. The results of such studies suggest that the phenomenon of the humanisation of robots can be viewed as a case of a more general phenomenon of animistic thinking.

People's tendency to animate their own creations has been known since antiquity. Ever since people started to create symbolic images of gods and spirits, they began to invest these images with thoughts and feelings similar to their own and perceive these images as if they were living creatures. One famous Greek myth tells the story of the sculptor Pygmalion who fell in love with the statue of a beautiful woman he had created. Aphrodite, the goddess of love, took pity on Pygmalion and gave life to the statue, which became Pygmalion's wife[29]. In ancient Rome, people used to have in their private homes small images of domestic deities – *lares* and *penates* – and in public places they erected statues of the main gods and goddesses, such as Jupiter, Juno and Minerva. Not surprisingly, monotheistic religions tried to uproot this tendency to animate images. The Bible says, "You shall not make for yourself an idol, or any likeness of what is in heaven above or on the earth beneath or in the water under the earth. You shall not worship them or serve them; for I, the LORD your God, am a jealous God…" (Exodus 20:4). Even more rigorous is Islam, which forbids the creation of images of sentient beings.

However, people's tendency to attribute spiritual qualities to their own creations is irrepressible; thus, in everyday Christian religious practices today, some icons are treated by believers as holy and having spiritual content to them. The humanisation of robots can be viewed as a modern twist of animistic thinking. Being a high-tech version of Pygmalion's statue, a robot is even more likely to be attributed with human qualities due to advances in prosthetics. While being still in its infancy, prosthetics nevertheless has achieved notable successes[30]. By reading intentionally induced signals coming from their user's nervous and muscular systems, biosensors relay this information to the controller (a device connecting the signals to the robotic limb), which then activates the artificial hand or leg. Some prostheses can be permanently attached to the body. In the future, prostheses could be created that assist not only the limbs, but also other vital organs, such as the heart, skin and liver. Extending this line of thinking leads to the notion of a *cyborg* – a being with both organic and biomechatronic body parts[31]. The notion of a cyborg aligns with the notion of a *biorobot* – a robot that emulates or simulates a living biological organism[32]. A synthesis of a cyborg and a biorobot could lead to a notion of an *android* – a humanoid robot that looks and acts like a human[33].

But here a problem arises: to what extent can a living human person endure the replacement of his or her organs with artificial ones before this 'Terminator' stops being a living organism and becomes a proper robot? The answer to this question goes beyond the technical capacities of bionics and enters the domain of metaphysics and philosophy. Where is this subtle border for the proportion between living and non-living organs in an organism, crossing which would turn a cyborg, which is predominantly a human being, into an android, which is predominantly a machine? Let us call this proportion '*the proportion of life*'. The fundamental difference between a cyborg and an android is that the android lacks the most important property of any living creature – the ability to independently generate subjective experience in the form of goals.

At first glance, we can make a robot generate independent goals by integrating into the machine a random number generator and a large number of possible strategies of action in a wide range of possible problems. Yet the difference between the goal-setting processes of a robot and those of a living organism would still be crucial. The living organism sets goals for itself not by the random selection of suitable strategies from an available range of options; *the living organism creates its goals sensibly, on the basis of subjective experience*. As argued earlier in this chapter, subjective experience is a magical phenomenon and cannot be reduced to physical processes, such as the functioning of the brain or signals coming from the environment. Subjective experience is a unique feature of a living entity that distinguishes a living creature from any, however complex, non-living structure, such as an android. The presence or absence of subjective experience sets the 'proportion of life' and defines the limits of prosthetics.

So, how can we determine the 'proportion of life'? The 'proportion of life' can hardly be established on the basis of pure theory, but it can be detected empirically by replacing the organs of a living organism with artificial organs. Perhaps not all human organs can be replaced by non-organic prostheses. For instance, it would hardly be possible to create a working prosthesis of a human brain. It is possible to create a non-organic model of a neuron that might imitate some of the neuron's functions. But the problem is that a neuron is not only a device for the transmission of information throughout the body via electric and chemical signals; the neuron is also a living cell. This means that a neuron, in some elementary form, possesses the ability to have some kind of subconscious subjective experience, and subjective experience cannot be modelled on a non-organic structure.

Of course, there is a chance that artificial organs will be grown from organic tissues. But then a person who gets these organic 'prostheses' will remain vulnerable to all of the limitations of organic life, including damage and death. In this 'organic' version of prosthetics, a person's life could be extended, but the person would never become a potentially immortal robot. Contrary to the ultimate expectations of transhumanism, and in spite of all the present and future successes of bionics, cybernetics and prosthetics, humans will remain humans – limited and mortal living beings, but beings with a divine gift: consciousness and subjective experience.

The impossibility of *The Matrix*

The Wachowskis' film *The Matrix* appeared late in the twentieth century, at the peak of 'computer euphoria'. Around a decade before the movie came out, the idea had emerged that it was possible to simulate the entire universe by converting every atom into a series of ones and zeros. Because a person is a part of the universe, in this cosmic program, every individual would occupy his or her humble place. "There is no way," American physicist Frank Tipler writes, "for the people inside this simulated universe to tell that they are merely simulated, that they are only a sequence of numbers being tossed around inside a computer and are not in fact real"[34], p. 181.

How do we know that we are real and not a simulation in some gigantic computer, Tipler asks? And he answers: we don't, but it doesn't matter. Likewise, it doesn't matter whether the universe is real or merely a simulation. All that matters is whether it is possible to create an abstract program that is capable of simulating the whole universe. There is a concept called 'data compression' in algorithmic information theory[35]. According to this concept, creating a computer program makes sense only if the program is shorter than a simple description in digital code of the process that this program codes. Accordingly, Tipler's question might be reworded as follows: could an algorithmic program code the whole universe, and if it could, then would this program be shorter than the description of the universe in digital code?

But algorithmic programs are created by people, and are pieces of knowledge. This fact brings us back to the problem of the relationships between knowledge and subjective experience. What comes first: knowledge about the structure of universe or subjective experience through which this universe is observed? If we accept the view that knowledge is a model of perceived phenomena and their interactions, then we first have to be able to perceive the phenomena before we can convert these phenomena into the symbolic reality of concepts, numbers and computer programs (see Figure 3.1). It becomes clear that computer programs of the universe could only be developed by an active agent who has subjective experience and is aware of his or her own existence.

So, who could be the programmer of the universe? That no intelligent life in the universe could be such a programmer is explained by Gödel's theorem. Simply put, according to Gödel's theorem of incompleteness, any encircled system of knowledge cannot explain itself without referring to something outside the circle – something you have to make assumption about but cannot prove[36]. For instance, it is impossible to decide whether the statement 'this sentence is false' is true or false while remaining within the rules of formal logic at the same time. The same applies to the universe. Gödel's theorem proves that it is impossible to simulate the universe from inside the universe. But is it possible to do so from the outside?

If the programmer is a living agent with subjective experience infinitely more powerful than human subjective experience, then the answer could be 'yes'. The Bible is written about exactly such a 'programmer'. But if the programmers were the robots from *The Matrix*, the answer has to be 'no'. Being non-living creatures, the robots could not possibly know what subjective experience is.

Not everyone will be persuaded by this argument. American physicist Michio Kaku writes about the possibility in the future of separating human consciousness from the human body and feeding it into a computer[37]. But before speaking about the future, it is useful to look back at the past, and looking at the past, we see that the first 'model' of our visible world was the invisible world of spirits (see Chapters 1 and 2). Computers could not invent the world of spirits because, being non-living things, they are immortal. Only living mortals who dream of an afterlife might need to create the world of spirits and benefit from its creation. This means that consciousness can only emerge in living entities. But this puts a fundamental limitation on the potential capacities of computer technologies. Even with all of the computing power of robots, one can't expect them to ever be able to simulate subjective experience.

Having said this, I don't mean to diminish the achievements of modern computer technologies in the domain of brain–computer interfaces. These achievements are fascinating. A functional magnetic resonance imaging (fMRI) scan can approximate which areas of the brain are associated with solving certain tasks. By scanning electric potentials of the brain, it is possible to teach a disabled person to 'mentally' control a wheelchair or a computer cursor. It is possible that, in the future, advanced methods of analysis of the brain's electric potentials might establish correlations between patterns of the brain's electric activity and the subjective images we see in our dreams. But all of these achievements won't change the fact that subjective experience and the electrical signals of the brain exist in different realms. There will always be a 'neutral strip' of the unknown between subjective experience and computer-simulated virtual reality. Rewording a known paradox, one might say that if human subjective experience were so simple that computers could simulate it, humans would not be so clever as to be able to create computers.

Conclusion

We started this chapter by asking the following questions: can robots simulate human subjective reality? Is it really the case that a person who has been plunged into a world of simulated subjective experience would be unable to tell this world from the real world? What is better: enjoying life in the illusory world or struggling for happiness in the true world full of hardships and austerity?

Our analysis discovered that subjective experience is a magical phenomenon of the 'emergence/vanishing' type. In the realm of subjective experience, the laws of magic hold sway. Subjective experience is a basis from which modern science and logical thinking grew; an implication of this fact is that subjective experience cannot be explained in terms of physical causality or formal logic. Subjective experiences can be studied, both objectively and 'from the inside', but studying subjective experiences require special methods that are different from the methods used in physics and physiology. The robots of *The Matrix*, created by humans on the basis of formal logic, would not be able to simulate subjective experience. The barrier that prevents robots from doing this is the necessity of being alive. Unlike with robots, under certain conditions neuroscientists can elicit simulated subjective experiences by tampering with a person's brain, but the person detects the difference between his or her authentic subjective experience and the simulated one.

What do we need to know this for? We need this so that we don't waste our time and effort on chasing unattainable goals, and so that we don't confuse studies of human subjective experiences with studies of inanimate objects in natural sciences. Or perhaps we need this for a better realisation of how unique and irreplaceable life and human beings are in this universe. Finally, we need this for an understanding that the explanatory power of the physical sciences, though enormous, still has its limits. However far into the brain, inside elementary particles of matter or into the depths of the universe science will go, it will not be able to escape the magic of subjective experience.

As for the question of whether it is better to enjoy life in the illusory world of simulated subjective reality (if some genius were ever able to simulate such a world) or to live a difficult life in the real world, I agree with the creators of *The Matrix*: this is a matter of personal preference.

Notes

1 https://en.wikipedia.org/wiki/The_Matrix
2 http://en.wikipedia.org/wiki/Huntington%27s_disease
3 Farah, M. (2005).
4 www.eurekalert.org/pub_releases/2006-10/jaaj-eoc092806.php
5 Selten, J. P., Cantor-Craae, E. & Kahn, R.S. (2007)
6 Caspi, A., Sugden, K., Moffitt, et al. (2003)
7 Stratton, G. (1896)
8 Gregory, R. L. (1970)
9 https://en.wikipedia.org/wiki/Neuroplasticity#Brain_training
10 Crick, F. (1994)
11 Eagleman, D. (2012)
12 http://en.wikipedia.org/wiki/Wilder_Penfield
13 Penfield, W. (1975)
14 https://en.wikibooks.org/wiki/Structural_Biochemistry/The_Oparin-Haldane_Hypothesis
15 https://en.wikipedia.org/wiki/Miller–Urey_experiment
16 Wells, J. (2012)
17 http://en.wikipedia.org/wiki/Coma#cite_note-18
18 Libet, B. (1999)
19 Wegner, D. (2002)
20 https://en.wikipedia.org/wiki/Stanford_prison_experiment
21 www.lancaster.ac.uk/staff/subbotsk/Shaping%20moral%20action.pdf
22 https://en.wikipedia.org/wiki/Milgram_experiment
23 http://en.wikipedia.org/wiki/History_of_neuroscience
24 Levy, D. (2007)
25 www.nbcnews.com/id/21271545/
26 http://articles.latimes.com/2007/nov/25/books/bk-lloyd25
27 www.washingtonpost.com/wp-dyn/content/article/2007/12/20/AR2007122002662.html
28 De Lange, C. (2014)
29 www.greekmyths-greekmythology.com/myth-of-pygmalion-and-galatea/

30 https://en.wikipedia.org/wiki/Prosthesis#Robotic_prostheses
31 https://en.wikipedia.org/wiki/Cyborg
32 https://en.wikipedia.org/wiki/Biorobotics
33 https://en.wikipedia.org/wiki/Android_(robot)
34 www.geocities.ws/theophysics/pdf/tipler-omega-point-as-eschaton.pdf
35 https://en.wikipedia.org/wiki/Data_compression
36 www.perrymarshall.com/articles/religion/godels-incompleteness-theorem/
37 Kaku, M. (2014)

PART II

The supernatural in science and religion

4

MIRACLES IN LAW

Magical underpinnings of the physical universe

Problem

Recalling my school lessons on physics, I see in my mind's eye the pictures of my school's physical laboratory, an induction coil, a device for the demonstration of electric discharge and a small cloud chamber in which alpha particles, like jets in the blue sky, left thin traces of steam. The names of Michael Faraday, Thomas Edison, Ernest Rutherford and other great scientists, who gave the world its modern physics, cross my mind. With all the diversity of trends in physics, physicists of the past were united in one thing: they knew how to ask nature questions and get answers. The theories they created were sometimes questionable, but an experiment always came to their aid. The experiment was a convincing judge: one can't argue with facts.

I did not become a physicist. However, I have maintained my interest in physics, trying to make up for any insufficiency of knowledge by reading books. Simultaneously, in the course of my career as an experimental psychologist, I developed an interest in magical thinking in modern people. To my surprise, I started to become aware of the fact that some theories in the books on physics for a general reader increasingly overlapped with magical phenomena. In the beginning, I explained this fact as inaccurate interpretations of new physical theories in the books written by science writers and journalists. But when established physicists started writing books for non-specialists and the overlap between physics and magic didn't disappear, my suspicion that magical phenomena had crept into modern physics grew stronger.

Of course, physicists will be quick to object and say that magic is the direct opposite of physical theories. Why? Because physical theories are based on facts that can be verified by experiments, and magic (e.g., astrology or palm-reading) is based not on verified facts, but on false beliefs. It is true that the laws of classical physics are immune to magical spells. Yet, in the domain of human consciousness and thinking, magical effects are real[1]. One important function of experiments in science is exactly to prevent magical phenomena, which happen in thinking or in the imagination, from invading scientific knowledge. But recently there have appeared theories in physics that are impossible to verify by experiments. One might ask: is it possible that, with experiment being impossible, there emerged a 'psychological wormhole' through which magical phenomena could trickle into physical knowledge? In this chapter, we will examine this possibility in the context of recent psychological studies on magical thinking and magical beliefs in modern people.

Magical thinking and the magical world

My interest in magical thinking emerged over 30 years ago when I observed children playing games of pretend. The feature that surprised me was how easy it was for a five-year-old 'player' to overcome the most insurmountable obstacles and solve the most difficult problems. "A tiger is chasing you, what are you going to do?" I ask a child, and the child answers, "I'll run away from it."

"And if there is a precipice in front of you?"
"I'll jump over."
"And if the precipice is very wide?"
"I will make a bridge and run over it."

In play, a child can pretend to be a bird, fly to another planet, read the mind of a wizard, erect or demolish a castle with the help of a magic wand. In other words, in play, the child 'thinks magically'. As stated in Chapter 1, magical thinking follows the laws of magic and embraces phenomena that do not conform to the laws of logic and science.

Indeed, when we do mathematical calculations or create a computer program, we rely on rational thinking. Rational thinking follows the laws of formal logic, where consequences follow from premises and the law of contradiction is observed. Thoughts are divided into steps, and each step follows from the previous one on the basis of a set of strict rules. In contrast, when we think magically, consequences don't need any premises, and two different objects or thoughts can merge into one. In magical thinking, thoughts follow each other not in a logical sequence, but by association. Images of objects can appear from nothing; this could be images of ordinary objects, but also bizarre combinations of known objects (e.g., a winged horse, a mermaid, a centaur) or objects we have never seen before (e.g., some objects portrayed in surrealist paintings). Magical objects can have unusual properties: demons that sort out fast molecules from slow ones; animals that can speak human languages; gods and spirits that can read human minds and be in several places simultaneously.

But isn't yesterday's magic today's science? As British science fiction writer Arthur C. Clark said, "Any sufficiently advanced technology is indistinguishable from magic" [1], p. 36. Indeed, just 300 years ago, such events as the instant transfer of a visual image over long distances and flying in the sky would be viewed as magical events, but today they are technological facts. Some scientists forecast that in the future even more advanced 'magical' devices will arrive, which may allow time travel and teleportation [2].

The parallel between magic and advanced technology is tempting, yet it is not entirely accurate. There are important differences between a magical action and a function of an advanced technical device. In the world of classical physics, everything – from an atom to a human being – is a complex mechanism. The objects that classical physics studies are soulless and mindless machines. By contrast, in the world of magic, all objects are animated; in this world, there is no difference between living and non-living things, there are no dead objects; every object, every entity, has a soul and a mind of its own. Every entity can be spoken with – all you need is to know the language. As argued in the Introduction, there is a crucial difference between magical communication and physical interaction – the same difference that distinguishes an invitation to sit down from making one sit down by brute force. Physical interaction is based on the four known fundamental forces – gravitational, electromagnetism and the strong nuclear and weak nuclear forces [3] – whereas magical communication, just like social communication, is based on semantics (see Table 0.3 in the Introduction). For instance, when we ask a person at a table to pass a saltcellar, we initiate the person's action, which cannot be reduced to the physical energy that we spend on speaking the words. This difference between magical and physical types of interactions results in the fact that, in the magical world, the laws of nature, which govern the world of technology, are invalid.

The laws of science and the laws of magic

One fundamental law of nature is *the law of identity*. According to this law, two objects or events are independent of each other unless a chain of physical causes continuously connects the objects. By this law, each object conserves its identity and is not influenced by other objects in the universe in any other way except through the aforementioned four fundamental physical forces. In contrast, in the magical world, two different objects can have a connection one with the other that cannot be causally explained. For example, damaging the body of Koschei (a character of Slavic folklore) cannot kill him. His soul is hidden inside a needle, which is in an egg, which is in a duck, which is in a hare, and so on. To kill Koschei, the hero must find and break the needle.

Linked to the law of identity is *the law of permanence*. This law requires that a complex object maintains its identity throughout a certain period of time and cannot instantly turn into another complex object. This law is challenged in the magical world, in which transmutation of one complex object into another (e.g., a person into an animal and vice versa) could happen.

Another law of nature is *the law of conservation*. According to this law, the amounts of energy, matter or momentum in the universe remain the same, whatever transformations happen to separate objects. Unlike in the world of nature, in the magical world, *things can appear from nothing or vanish without a trace* ('emergence/ vanishing' magical phenomena). A ghost or a genie can suddenly materialise in front of a person, a cat can vanish into nothing with the last thing visible being its grin; UFOs, the Sasquatch, the Loch Ness monster and other enigmatic, uncatch- able entities can emerge from nowhere and disappear into thin air.

Finally, one more law that is crucial for science is *the law of independence*: the objects and processes of nature are independent of human thinking. For instance, the trees that obstruct our view of the ocean remain there however hard we wish them to go, and the law of gravitation doesn't care whether we like it not. Similarly, a person can't access another person's mind, except via communication through physical media (the air or an electromagnetic field). This fundamental assumption underlies the principle of objectivity of scientific research. In research, experimental facts are independent of how we want them to be; experimental results define the explanations, not the other way around. In contrast, in the magical world, the mind can change matter (direct Self over matter magic) or another mind (direct Self over mind magic) without connecting to the matter or the mind via the existing physical forces. It is easy to see the difference between magical 'direct Self over matter/mind' actions and the actions of modern devices that react to electromag- netic waves of the brain (e.g., motor prosthetics or brain–computer interfaces[4]). The principle on which neuroprosthetics are based looks magical, but this resemblance is only superficial. Neuroprosthetics do not decode the meaning of a thought (e.g., "I want to move my hand"), but react to the electric signals or muscle contractions that are induced by the thought, thus converting these signals into movement (e.g., switching on the motor of an artificial limb or shifting a cursor on a computer screen). In essence, neuroprosthetics are highly advanced and sophisticated versions of a remote control. On a remote control, we press buttons with a finger, whereas on a neuroprosthetic device, a disabled person 'presses the buttons' by intentionally producing appropriate electric impulses via thinking about making a certain action. This kind of interaction is really magical only at the person's end (a thought or wish affecting the pattern of electric signals in the brain and nervers; see Table 0.3), whereas the device automatically reacts to the physical force – the electromagnetic impulses or muscle contractions that correlate with the thought or the wish (see Chapter 3 for more on this). By contrast, at the receiving end of a magical incan- tation or a prayer, there is another mind – a spirit inside the physical object, the mind of another person or the mind of God – which considers the plea and makes a decision about whether to grant or reject it.

Let us make a brief summary of the relations between the laws and phe- nomena of magic and the laws of science. As argued in the Introduction, magical thinking follows *the law of sympathy*, according to which two objects that resemble one another (e.g., a person and his or her photograph) are connected and share a common fate: if the person's photo is hurt, the person will suffer as well, even if the person is in a different part of the globe. Another law of magic is *the law of contagion*: objects that once were one (e.g., a person and a bunch of the person's hair) maintain their link forever; if anything happens to one of the objects (e.g., the bunch of hair is burned), then the same thing will happen to the other (e.g., the person will get ill).

Along with the *laws of magical thinking*, which are similar to the laws of nature in that they work universally and continuously, there are also *magical phenomena* – events that violate the laws of science yet cannot be deduced from the laws of magic either. Magical phenomena don't have the universal status of laws. *Whereas the laws of magical thinking are confined to the domain of mental processes, such as perception, thinking and imagination, some magical phenomena can occasionally occur in the real, physical world.*

One of these phenomena is *participation*, according to which two objects that are entirely different from each other, don't resemble each other and have never been in contact could nevertheless share a common essence or soul. French anthropologist Lucien Lévy-Bruhl illustrated the phenomenon of participation with the example of Bororo tribe of central Brazil, who mystically identified themselves with arara parrots, which they viewed as their totem. It was not that the Bororo did not see a difference between a person's physical appearance and that of a parrot; yet, with all the physical differences between a parrot and a person, the Bororo maintained that a spirit of an arara lives in a person, and so in each Bororo there is a part of an arara, and vice versa[5]. The second magical phenomenon is *emergence/vanishing*: things can emerge from nothing and can vanish into nothing. Manifestations of this phenomenon are random events and creative thinking. In the magical world, nothingness is not a sheer absence of something, but is a productive 'black hole' that can generate random events and creative ideas; it can also engulf events, objects and ideas, which then disappear without a trace. The third magical phenomenon is direct '*Self over matter/mind*': Self (mind) directly affects physical objects and processes or another person's mind; for example, the gods create the world, a prayer to the gods helps with hunting or growing crops, or a person's effort of will affects the working of a physical device. The difference between direct *Self over matter/mind* and *emergence/vanishing* magical phenomena is in that direct Self over matter/mind involves an active agent, whereas emergence/vanishing doesn't. Finally, the fourth magical phenomenon is *transmutation* – the sudden changing of one complex object into a different complex object (e.g., a person changing into an animal, and vice versa).

The laws of magical thinking, magical phenomena and the laws of science are brought together in Table 4.1.

Because the laws of magical thinking existed long before science, these laws could not develop by negation of the laws of nature; rather, science built its laws by rejecting the laws of magical thinking and prohibiting magical phenomena. Nevertheless, the world of nature is not separated from the world of magic by an impregnable wall. As we will see in this and subsequent chapters, *certain magical phenomena (such as participation and direct Self over matter) can be observed in the real, physical world.*

Similarly, the imaginary magical world is not isolated from the natural world, but is a mixture of ordinary and magical things and events. For example, in Alexander Pushkin's fairy tale *Ruslan and Lyudmila*[6], Chernomor the Wizard in some ways

TABLE 4.1 The laws and phenomena of magic and the laws of science

Laws and phenomena of magic	Laws of science opposing the laws and phenomena of magic	Phenomena allowed by the laws of magic and prohibited by the laws of science
The law of sympathy (two objects that resemble one another are connected and share a common fate)	Identity (two objects or events are independent of one another unless they are continuously linked by one of the four physical forces)	Affecting a person's photograph will affect the person, immediately and at an indefinite distance
The law of contagion (two objects that once were in contact maintain their link forever)	Identity (two objects or events are independent of one another unless they are continuously linked by one of the four physical forces)	Affecting a person's bunch of hair will affect the person, immediately and at an indefinite distance

A person's qualities pass to another person via personal things |
| The phenomenon of participation (two objects that are entirely different from one another, don't resemble one another and have never been in contact can still share a common essence or soul) | Identity (two objects or events are independent of one another unless they are continuously linked by one of the four physical forces) | A tribe and its totem (e.g., an animal) share the same invisible essence or soul

A person's fate is connected to a certain planet

Sacrificial bread and wine transubstantiate into the body and blood of Christ |
| The phenomenon of emergence/vanishing (things can emerge from nothing and vanish into nothing) | Conservation (the amounts of energy, matter or momentum in the universe remain the same) | Random events appear from nothing

Creative ideas appear from nothing

Ghosts and apparitions emerge and vanish |
| The phenomenon of direct Self over matter/mind (an effort of will or desire can directly change physical objects and processes and pass information to other minds) | Independence (the objects of nature are independent of human thinking and individual minds are not directly linked to one another) | Prayer changes physical matter

A person affects external physical processes or passes information to another person without connecting to the processes or the other person via a physical medium

Mental processes directly affect physical processes (patterns of electrical activity in the brain, behaviour of a quantum object) |
| The phenomenon of transmutation (complex entities instantly turn into other complex entities) | Permanence (one complex object cannot instantly turn into another complex object) | A person turns into an animal and vice versa |

is an ordinary person: he walks in his garden, falls in love with beautiful girls and even suffers from a sexual dysfunction. However, he also has a magical beard, which allows him to fly in the air and perform miracles. An equally odd combination of the ordinary and the magical, the possible and the impossible we see in the myths of almost every nation. Because these 'games with magic' are confined to the realm of our fantasy, they peacefully coexist with our belief in science. In fact, most educated adults in industrial cultures today, while enjoying magical thinking, are convinced that they don't believe in magic.

However, psychological studies of recent decades have revealed that despite their conscious disbelief in magic, in their subconscious, educated adults continue to believe in the supernatural. In one of these studies, British university graduates and undergraduates were shown an apparently 'magical' effect – a square, plastic card becoming badly damaged in an empty box after a magic spell was cast on the box[7]. The participants were then tested in (a) the low-risk condition, where their driver's licenses were at risk of destruction by a magic spell, or (b) the high-risk condition, where the participants' own hands were the objects at risk of being damaged as a result of the magic spell. The results showed that, in the low-risk condition, only 12% of participants prohibited the magical spell; however, in the high-risk condition, 50% of participants asked the experimenter not to repeat the spell and justified their decision by admitting that the magic spell might indeed damage their hands. Other studies also indicate that, in certain circumstances, modern educated adults exhibit their belief in magic, in both their intuitive reactions[8] and conscious behaviour[9].

But if educated adults, when placed in a certain context, are prepared to accept that physical objects succumb to the laws of magic in a laboratory setting, what can prevent people from believing that certain magical phenomena could occur in nature as well? Is it possible that even scientists – the creators of theories in physics – could admit magical phenomena into the world of nature, especially if mathematical calculations bring them to the conclusion that magical events are real? After all, physicists are people. Like all people, they are locked in their shell of consciousness and have to conform to all the magical phenomena that consciousness entails.

Science or quasi-religion?

Physical science, due to its technical achievements, won such great respect from modern society that, for a layman, physicists' statements became almost sacred. For scientists who are not physicists, the trust in modern physics is based on their respect for precise and elaborated experimental procedures that provide empirical support for physical theories. But what about the latest physical theories that cannot be tested by experiments? In what way are these theories different from the stories of the origins of humanity and nature that are told in the Bible? These theories are based on mathematical equations and post-hoc evidence, but mathematics is a science about numbers, not about the laws of nature, and post-hoc evidence (e.g., the expansion of the universe) is based on interpretations rather than on empirical testing.

The first time this question crossed my mind was many years ago when I began reading books about the theory of relativity. As is well known, Einstein's special relativity theory is based on Galileo's principle of relativity, which states that the laws of physics are identical in systems moving in a straight line, one at a constant velocity in relation to the other, so that one cannot conduct any physical experiment capable of indicating if the body is immobile or in motion. Systems that move in a straight line with constant speed relative to each other are called inertial frames of reference, or 'inertial frames' for short. Galileo invented transformations that can help compare two inertial frames. Einstein applied Galileo's principle to the laws of optics and electrodynamics and drew a conclusion that an absolute reference frame doesn't exist[10].

The question I pondered was how the concept of physical inertial frames of reference related to the imaginary frame of reference that exists in the head of a theorist. Indeed, suppose I am thinking on the popular problem of whether a certain event (e.g., lightning) occurs simultaneously or at different times in two different reference frames (e.g., a train station and a train that is passing by). Independently of whether I am standing on the railway platform and thinking about the train that is passing by or sitting in the moving train and thinking about the platform, I have both of these frames of reference in my head. It turns out that the statement on the equivalence of the laws of nature in inertial frames only makes sense if I tacitly allow for the existence of my own 'personal' frame of reference, from which I can compare the two inertial frames and make the inference about the impossibility of distinguishing between them by any physical experiment. Embedded within his or her 'personal' reference frame, an observer can theorise about 'jumping' from the platform onto the train and back and conclude that the laws of nature in both are identical. What one can't do, however, is jump out of one's own mind and ponder the laws of nature while being 'outside' oneself at the same time. Subjective experience is the absolute frame of reference, which is always with us (see Chapter 3). *Interestingly, most physicists tend to ignore this psychological frame of reference as though it doesn't exist.*

Reading various books and roaming the Internet for an answer, I discovered that criticising relativity theory is considered to be in bad taste, similar to trying to invent a perpetual motion machine. Most criticism that exists is online[11,12]. The impression is that the theory of relativity became a kind of 'quasi-religion' of modern physical science, like Marxism was in the former Soviet Union. Of course, calling Einstein's theory a 'quasi-religion' is only a metaphor. However, regarding certain modern physical theories, this metaphor is close to the truth. Indeed, today many scientists have started to contemplate theories that are impossible to verify empirically: 'string theory', 'M-theory', and the 'theory of a multiverse' are only a few among this class of theories[13,14]. On the basis of these theories, it is hard to expect new experimental discoveries any time soon. Of course, rumour has it that the renowned British physicist Lord Kelvin once assumed that there was nothing left to discover in physics, after which a flood of major discoveries followed[15]. And yet, it seems that Lord Kelvin, if he said that, was not entirely wrong. Physics' explanatory power, like that of any scientific discipline, will inevitably come to its limits.

It is possible that in certain areas physics has already met its limits to explain reality. There is also the opinion that physics and some other sciences became victims of their own success. The higher the tempo of discoveries in science, the sooner the science runs into a certain insurmountable explanatory boundary[16]. Physics can never give ultimate answers to the questions of what physical reality is 'really like', how the universe emerged, what the fate of the universe is beyond a few billion years, why the fundamental physical constants (e.g., the speed of light, the gravitational constant, the Plank constant) are so finely tuned to each other that they make life and a human being possible, what the role of a human being is in the universe and many others. It is unlikely that biological science will ever be able to create living things from non-living ones, or even completely understand the mechanism of how an individual organism is built. We cannot expect psychology to fully explain consciousness, or cybernetics to prove that machines are capable of conscious thinking.

When physics as an empirical science comes to a threshold beyond which facts have to give way to beliefs, the belief in magic, which thus far has been lurking in physicists' subconscious, can trickle into physical theories. And the situation appears when it is hard to understand which of the theoretical constructions is still science and which is already magic.

So, what makes this mind-boggling interaction between theoretical constructs of modern physics and magical phenomena possible? It appears that the causes of this convergence are hidden in the structure of human thinking and in the subconscious belief of modern people in the supernatural.

Magic as a 'midwife' of physics

The history of science suggests that magical thinking and scientific thinking are not incompatible intellectual processes, but instead team up in the search for the truth. As some authors have pointed out, science grew out of magic: astrology gave rise to astronomy, alchemy gave birth to chemistry and modern mathematics originated from numerology[17]. Another branch of culture that grew out of magic was modern religion. In prehistoric times, magic and religion were the same. Finally, human language and poetry emerged form magic as well. The first word was a magic spell, the first sign was a magical symbol and the first poetry was a way of speaking with the gods. This is particularly evident when one looks at Palaeolithic cave paintings, Egyptian hieroglyphs and Greek mythology[18]. Religion and science grew out of magic by negating it, and language and poetry by taking from magic and reworking magical rituals into symbols and metaphors. That could be the reason why religion and science are hostile towards magic, whereas human language and poetry openly acknowledge their kinship with the supernatural. As stated in the Introduction, communication through language is based not on physical causality, but on a magical one. In essence, a magical ritual is the forerunner of the whole of culture[19,20]. However, different branches of culture separated from magic at different times.

Unlike religion, which has fought with magic since biblical times, in the age of Isaac Newton (1642–1727), physical science was not yet in opposition to magic, but peacefully coexisted with it. In the seventeenth century, physical science was the occupation of few, and the limits of science were more visible than they are today. Science did not aspire to understand the human mind, which remained in the department of religion. Newton was not just a mathematician and a physicist, but also an alchemist, who tried to find the 'philosopher's stone' that could transmute common metals into gold. According to the Alexandrian cultural tradition (second to fourth centuries CE), there was a magical kinship (sympathy) between certain metals and the seven planets of classical astrology. The sun was 'the ruler' of gold, the moon of silver, Venus of copper, Mars of iron, Jupiter of tin, Mercury of mercury and Saturn of lead[21]. In Newton's time, an impermeable barrier did not yet separate astronomy and mathematics from astrology and alchemy. Indeed, half a century before Newton took a post in Trinity College of Cambridge University, a member of this college's council was John Dee (1527–1609), a famous mathematician, astronomer, alchemist and astrologist.

John Dee's contemporary, Italian Dominican friar and philosopher Giordano Bruno (1548–1600), was not only a mathematician and an advocate for Copernicus' heliocentric system, but also studied magic and astrology. Bruno defined magic as knowledge of the science of nature[22]. Conflating science and magic was not unusual in the time of the Renaissance, because both magic and science sought an answer to the same question of how to gain control over the universe. In a time when religion had long been persecuting magic (one of the accusations of the Holy Inquisition against Bruno was that he had been involved with magic), science still did not cut the 'umbilical cord' that connected it to magical thinking. Perhaps it was the magical kinship between the planets and also between the planets and metals that brought Newton to the idea of gravitation – the enigmatic, invisible force that works instantly and at a distance. Gravitational force seemed to Newton as incomprehensible as astrology's 'force of sympathy', which connected planets and people. Only in the beginning of the twentieth century did Einstein, with his general relativity theory, manage to explain gravitation through the curvature of space. So it appears that the belief in magic opened the gate to classical physics.

The question arises as to why physical scientists as outstanding as Isaac Newton did not see a contradiction in studying magic and physics simultaneously. Was it because magical and scientific thinking need not compete one with the other, but could cooperate in a search for the truth?

The image of 'a rider and a horse': scientific thinking as 'taming the magical'

Sigmund Freud used the image of 'a rider and a horse' for describing the relationships between the conscious 'I' – the holder of one's will and rational thinking – and the subconscious 'Id' – the holder of drives and passions[23]. The same image comes to mind when we think about the relationships between magical thinking and scientific thinking. According to psychoanalysis, the subconscious is not only a source of drives, but also of creativity. *Because the realm of subconscious thinking is not strictly constrained by formal logic, the laws of nature and social stereotypes, new and unexpected combinations of images and ideas can emerge in this realm.* When these unusual combinations ascend into the realm of reflective consciousness, some of them are rejected by critical thinking (censorship), while others can give rise to new scientific theories. In essence, the relationships *between subconscious and conscious types of thinking* describe the relationships between *magical and scientific types of thinking.*

Indeed, as discussed above, the laws of magical sympathy and the phenomenon of magical participation connect two independent objects that objectively have no link to one another. In a similar way, symbols in our thinking work through participation. For example, when in a war movie that is placed in a certain socio-historical context we see a knife piercing into a tree trunk, we understand that war broke out. When in the same movie we see a dove descending onto a broken and rusty tank, we know that peace has come. The knife and the dove can be parts of both war and peace, but when they are put in specific contexts they symbolise either war or peace by embracing war or peace as a whole, with all the armies and states involved. According to Freud and some modern theorists[24], our subconscious thinking operates not through concepts, but through symbols. But symbols are also the language of magical thinking. As mentioned above, being free from stiff limitations of formal logic, magical thinking can generate original combinations of thoughts and thus play an important role in the search for scientific truth.

In point of fact, some scientists have explicitly acknowledged that magical thinking helped them to come up with new ideas, either through subconscious processes or via consciously designed thought experiments. For example, German chemist August Kekulé, who discovered the ring structure of a benzene molecule, confessed that right before this idea crossed his mind, he had dozed off in a chair and had a daydream of a *serpent* eating its own tail (this is an ancient symbol known as the Ouroboros)[25]. To illustrate the Second Law of Thermodynamics, British physicist James Clerk Maxwell used the image of a *demon* who violated that law by sorting out fast molecules from slow ones and thus decreasing entropy in a closed physical system[26]. Einstein mentioned that the idea of special relativity theory entered his mind for the first time at the moment when he *imagined himself* sitting on the end of a beam of light[27].

These revelations support the suggestion that a scientific discovery occurs not as an inductive generalisation of known facts, but as a 'spark' of magical thinking. New insights reveal themselves unexpectedly through magical participation, which links together totally different images (e.g., the structure of a molecule and a snake eating its tail) in a way that is impossible to rationally explain. It is only after a discovery is made and supported by experimental facts that our rational mind brings the succession of events 'in order', presenting the discovered law of nature as a logical conclusion from earlier established facts. In reality, as Einstein wrote, "There is no logical path to these laws; only intuition, resting on sympathetic understanding of experience, can reach them"[28]. In other words, magical thinking supplies rational thinking with new 'theories in the making', and rational thinking sifts these theories through the filter of experience. At the same time, by backdating, rational thinking tries to 'fix' the discovery-making process in order to exclude magical thinking from the picture.

Whereas some scientists, like Einstein, recognised the role of magical thinking in the scientific creative process, they nevertheless refused to admit that magical phenomena participate in the order of nature. Einstein's maxim "God doesn't play dice" prohibits random processes from participating in the fundamental laws of physics[29]. Indeed, by definition, a random event is in principle unpredictable and undetermined by physical causes. From the perspective taken in this book, this makes a random event *a magical event* of the 'emergence/vanishing' type (see Tables 0.3 and 4.1). That is why in classical physics scientists don't allow random events to intervene in fundamental laws of nature. We call the result of a dice roll a random event, but only mean that the randomness comes from our inability to account for atmospheric noise and all the tiny variations in the effort of a hand throwing the dice. But that doesn't mean that this result is random in the fundamental sense of the word (i.e., it cannot be causally explained 'in principle'). If we had a supercomputer that could take into account all the forces that affect this individual dice throw, then we would be able to exactly predict the result. As it stands, a 'random' event in everyday life is only 'pseudo-random'.

Just as it prohibits randomness, classical physics also forbids a direct effect of the Self on the results of experiments. Of course, an experimenter may affect experimental results in many ways, but these 'experimental biases' only happen due to imperfections of the experimental procedure. What is impossible is the magical direct 'Self over matter' effect (e.g., affecting the result of throwing a coin just by thinking hard about the desired outcome). Indeed, if the experimenter's mind could directly affect the result of an experiment, this would undermine the objectivity principle, which is fundamental for science[30]. However, the emergence of quantum mechanics in the twentieth century shattered the aforementioned prohibitions, thus opening the gate for magical effects to enter physical reality. It turned out that randomness is an unavoidable element of quantum processes, and the observer's consciousness plays a definitive role in what the observed physical reality is.

Participatory universe: the magical properties of physical reality

The magical law of contagion was coined at the beginning of the twentieth century by British anthropologist James George Frazer[31]. Recall that contagion is a magical bond between things that once were in physical contact but then separated one from another. This bond works instantly and doesn't depend on the distance between the bonded objects.

Similar invisible bonds and connections can exist between the quantum objects of modern physics. The most close to 'magical contagion' is 'quantum entanglement'.

Quantum entanglement is a physical phenomenon that occurs when a pair of quantum objects is generated that interact one with the other in a way such that the quantum state (e.g., spin) of each particle cannot be described independently of the other. If something happens with one of the entangled objects, then something also happens to the object's entangled pair, even if arbitrarily large distances separate the pair. It thus appears that one object of an entangled pair 'knows' what measurement has been performed on the other, and with what outcome, even though there is no medium between the objects to pass this information through. It is no wonder that Einstein, Podolsky and Rosen, who first described this phenomenon as a thought experiment, considered it to be impossible because it violated the principle of physical causality[32]. Yet later this counterintuitive prediction was verified in experiments[33]. It turned out that information instantly and without any medium known to physics gets from one entangled quantum object to the other. Some physicists argue that entanglement cannot be used for the faster-than-light transmission of meaningful information[34]. Nevertheless, the fact remains that some interaction between physical objects can happen in an instant, even if the objects are separated by a distance the size of the universe. In the mind of most adults today, only gods can be present in many places at once, and for children Santa Claus has the same ability. Clearly, entanglement is not a cause–effect connection, because a cause precedes its effect in time. Quantum entanglement is a magical contagion link between two physical events, or, as Einstein once called it, a "spooky action at a distance." There are interesting attempts to explain quantum entanglement by retrocausality. This assumption suggests that a measurement made by a scientist on one of the entangled particles can instantly change the properties of the particle in the past, when this particle was in close contact with the other entangled particle. The signal then could travel forward in time to the other entangled particle[35]. However, the retrocausality hypothesis simply replaces one magical effect (entanglement between two events in space) with another (entanglement between two events in time)

Another magical phenomenon is the participation between an observer of quantum phenomena and the phenomena being observed. According to the Copenhagen interpretation of quantum mechanics[36], no quantum object (e.g., a photon) exists as a certain definite entity prior to being measured. Prior to an act of measurement, the object exists in the 'superposition state' – an indefinite, probabilistic state in which it is impossible to say whether the object is a particle or a wave. It is the act of measurement that defines in what capacity – of a particle or a wave – the object reveals itself. This means that an act of observation doesn't simply reflect reality in the way a mirror reflects an object, but participates in what this reality becomes, influencing reality. In other words, there is a bond between an act of observation and the quantum entity being observed. But this bond is not of a cause–effect type, in which one physical object affects another physical object through one of the four known fundamental forces – electromagnetism, the weak and strong nuclear forces and gravitation. The bond between an observer and a quantum object is a magical bond in which two fundamentally different realities interact: mental (the observer's conscious decision to apply this or another way of measurement) and physical (a quantum object). A cause–effect type of interaction between two quantum objects conforms to the law of energy conservation (e.g., a collision between an electron and a positron creates two gamma ray photons). For a magical type of interaction, the law of energy conservation is irrelevant, since in this interaction a quantum object interacts with the observer's subjective experience, which has no physical properties. As a result of such an interaction, the quantum object doesn't turn into another quantum object, but adopts a certain form – of a particle or a wave. This phenomenon shows that the *subjective reality of the observer is not a mirror that simply reflects the quantum event, but a medium that interacts with this event and affects this event during the interaction.*

Indeed, we can only interact with external objects through our senses, and our senses perceive the object via certain media – air, water, glass or electromagnetic fields – and these media leave their marks on the resulting image of the object. For instance, seen underwater an object looks bigger than if it is seen through air, and white light can produce a spectrum of colours if seen through a glass prism. In other words, the properties of the medium become an inseparable part of the final image of the object. Summarising this, we could say that the medium has a certain 'viscosity', which brings changes in the image of the original object. Now, we also know that subjective reality is a magical medium that is given to humans for observing the things-in-themselves. Observers cannot observe a quantum object as a thing-in-itself; they can only observe the object clad in subjective experience. The question is whether subjective experience, like any medium, also possesses 'viscosity' of some kind or it is entirely transparent and exerts no resistance whatsoever. *If subjective experience were entirely transparent, it could not influence the collapse of the wave function.* The empirical fact that the way observation is made affects the result of the observation shows that subjective experience is a medium that does leave its mark on quantum objects. This is what participation between the observer's consciousness and physical reality means. *Exactly how human subjective experience, which is a non-physical entity, can directly affect physical objects we cannot possibly know, but the fact that it can is always in front of us in the form of our ability to intentionally move our bodies.*

The participation of a conscious observer in shaping physical reality can be illustrated by the famous 'double slit experiment'[37]. In one version of this experiment, a laser, which is a coherent light source, illuminates a plate that has two parallel identical slits cut into it. For simplicity's sake, let us call them Slit A and Slit B. On the screen behind the plate, the light beam leaves a trace. Sending photons one at a time shows that each individual photon passes through only one of the slits, and not through both as would be expected from a wave, producing a single spot on the screen. Thus far each photon behaves as a single particle of matter. However, when the number of spots accumulates, they build bright and dark bands on the screen, which is an indication of interference – a phenomenon that is a property of waves, but not of particles. This proves that a photon can be both a particle, which has a fixed localisation in space, and a wave, which has no such localisation – a property called wave–particle duality. Electrons and other elementary units of matter behave in a similar way[38]. Although wave–particle duality violates the properties of an object of classical physics, so far the observer registers the result without affecting it.

However, a modified version of this experiment can show how an act of observation affects the process of the photon taking a form of either a particle or a wave[39]. A special prism converts every photon emitted by the laser into two entangled photons of a lower frequency, which then are focused into two separate beams of light; for convenience, let us name these beams the upper and the lower ones. Next, one of the beams – the lower one – goes through the double slit plate, and the detector screen at the other side registers the expected interference pattern. At this stage, a new element is introduced to the experiment: a special device 'marks' the photons by producing clockwise circular polarisation of the photons that went through Slit A and anticlockwise circular polarisation of photons that went through Slit B. As a result, an observer could know which path each individual photon of the lower beam took. After this, the interference pattern on the detector screen disappears. Finally, another device gives the photons of the upper beam a linear polarisation that makes the photons indistinguishable from one another. By entanglement, this makes the lower beam's photons indistinguishable from one another as well. This erases the effect of distinctive circular polarisation of the lower beam, which now becomes a mix of clockwise- and anticlockwise-polarised photons. The photons of the lower beam are no longer marked and the detector screen of the lower beam can no longer determine through which slit each individual photon passed. As a result, the interference pattern on the detector screen reappears. The conclusion is clear: when the observer could know which path in space had been taken by each photon, the photons behaved as particles, and when the observer couldn't know 'which path' had been taken, the same photons behaved as waves. *It looks as though the observer's knowledge about the quantum object's path 'condenses' it into a distinct object – a particle, whereas when the knowledge about 'which path' is absent, the quantum object 'spreads' in space as a wave.*

At first glance, this effect of the observer on the observed is purely mechanical and is not a direct effect of human Self on matter. The way a quantum object collapses depends on our choice of measuring device and not on our will or desire. In this regard, the results of the double slit experiment are no different from the results of any observation in which physical devices are used. Thus, the same physical object looks different when seen through lenses of different magnification capacity, through different colour filters or through different media. There is, however, an important difference between observation of macro objects and observation of a quantum object. *Observing a macro object can change its appearance, but not its essence.* For example, we can use different methods to determine the object's molecular composition, chemical properties, temperature or weight, and the results will be the same independently of the methods; the results will also be independent of whether we see the object as big (when close to us) or small (when at a distance from us), red (under red light) or green (under green light). In contrast, the way we observe a quantum object defines the object's essence – to be either a particle or a wave. This specific feature of quantum measurement allows us to say that *in the quantum world, measurement involves a direct effect of a conscious observer on the quantum object: the observer's decision to ask nature a question, through his or her choice of measuring device, defines the fundamental properties of the natural object – being either a particle or a wave.*

One might ask what all of this has to do with consciousness. Indeed, photons and electrons are being registered not directly by our consciousness, but by devices. But this question is wrongly put. Our consciousness can't directly register not only photons and electrons, but also any object at all. In order to be able to see something, taste something or hear something we need special devices: the retinas in our eyes, taste buds in our mouth or eardrums in our ears. Such things as 'redness', 'saltiness' or 'loudness' do not exist in nature; what exists in nature are electromagnetic waves of a certain length and frequency, the molecular structure of a grain of salt and the amplitude of a sound wave's vibration (see Chapters 3 for more on this). It is our perceptual organs (eyes, mouth, ears) and brain that make conversion of these natural structures into subjective experience – colour, taste, sound and a geometrical pattern – possible. Devices that allow us to 'see' photons and electrons are our 'artificial eyes' that help to convert some invisible reality into the reality that is accessible to our perception – traces of photons or electrons on a detector screen. In other words, *physical devices are an extension of our sense organs through which our consciousness can see the world; therefore, these devices are part of our consciousness.* Thus, in the above double slit experiment, *detecting the photons' paths through clockwise or anticlockwise polarisation didn't change the photons' physical ability to produce interference pattern on their own. All it changed was adding the possibility of observer's knowledge about the photons' paths, and this possibility was sufficient to eliminate the interference. This makes the observer's effect on the collapse of the quantum object a direct Self over matter magical phenomenon.*

An interesting twist to this experiment is an experiment in which the decision of whether or not to erase the information about 'which path' is made after the photons that went through the slits in the slate reached the detector screen[40]. This is achieved by increasing the distance between the source of light and the screen that detects the upper beam's photons in such a way that the lower beam's photons reach their detector screen before the photons of the upper beam. When the upper beam's photons (which have a distinct linear polarization) reach their detector screen unhindered, everything goes as expected: marking the lower beam's photons' paths through the double-slit plate via circular polarization eliminates the interference pattern, like in the previously described experiment. But now a new feature is introduced: when the lower beam's photons have already 'collapsed' on their detector screen but the upper beam's photons are still on their way toward their detector screen, a special device makes the upper beam's photons indistinguishable from each other by mixing their linear polarization characteristics. By entanglement, this erases the information of 'which path' the lower beam's photons went through, and the interference pattern on the detector screen for the lower beam's photons reappears again. The amazing feature of this experiment is the very fact that mixing the upper beam's photons (and, therefore, erasing the information 'which path' in their lower twins) was done *after the lower twins had already reached the detector-screen.* It appears that the lower beam's photons 'knew in advance' what would happen to their entangled upper beam's twins, and thus 'understood' that the observer would no longer be able to learn through which slit each of the photons would go.

To summarise, it appears that the shape of a quantum particle can be determined by the observer's expectation alone. It is not that we make the photon abandon its wave-like behaviour by applying a physical force to it. No, just our knowledge of which path the photon would go if it were a particle and not a wave makes the photon behave as a particle. Similar direct effects of Self over matter occur when we voluntarily move our body, or when a disabled person's desire to move his or her absent limb changes patterns of electric activity in the person's brain and motor neurons.

These experiments demonstrate that human subjective reality is a medium that exerts resistance on things-in-themselves. The experiments show that, when a quantum object's path is not known, the quantum object tends to collapse as a wave. This suggests that collapsing as a wave is a 'normal' behaviour of a quantum object when its path cannot be detected; if this is the case, then *the object collapsing as a particle can be interpreted as a regular distortion brought about in the object's normal behaviour by the fact of detecting the object's path*. Through this distortion, the viscosity of human mental reality is revealed – a viscosity that could not be revealed when we observe macro objects due to the nearly absolute 'transparency' of mental reality as a medium. As the last of the above experiments suggests, this viscosity works not only in space, but also in time, through entanglement.

So far, so good. We established that magical thinking could play a vital role in scientific discoveries. We also revealed that in quantum physics, magical phenomena, such as magical contagion and direct Self over matter, can occur not only within magical thinking, but also in physical reality itself. The evidence that upgraded magical phenomena from being mental events to the status of physical events was obtained through rigorous experiments. The belief in the supernatural that modern educated people implicitly hold didn't play a role in these developments.

Yet the same experiments that demonstrated the definitive role of an observer in shaping the states of quantum objects brought physicists to the threshold, beyond which lies the realm of reality where experiments become inconceivable. In this hypothetical realm, only mathematics and theoretical reasoning hold sway. And where the empirical verification of theories is impossible, there appears a possibility for the implicit belief in magic to sneak into the body of physical knowledge.

Crossing the threshold: from the participatory universe to the multiverse, and further along the line

One influential theory of modern physics that is difficult to distinguish from pure magic is the so-called 'many-worlds interpretation' (MWI) in quantum mechanics. According to this theory, put forward by American physicist Hugh Everett[41], any measurement of a quantum object, such as the fixation of a photon on a photographic plate as a spot of light (usually called 'the wave function collapse'), creates not a single version of this object, as the Copenhagen interpretation of quantum mechanics insists, but splits the universe into the version that we see and the other possible versions, which stay invisible to us. Everett called his interpretation 'the universal wave function interpretation', and American physicist Bryce DeWitt later renamed this theory the MWI[42]. DeWitt elaborated that by every quantum measurement an observer unintentionally creates multiple universes–worlds, each including a copy of the observer inside it. It is hard to avoid a conclusion that in this interpretation of quantum measurement *an observer is a creator of worlds, with the only difference that God of the Bible created the universe willingly by inspiration, whereas the 'quantum god' manufactures universes unintentionally.* To believe or not to believe in this theory is a matter of preference. The question that some scientists ask is whether this theory could be tested in experiments[43]. According to cosmologist Max Tegmark it could, but only at the cost of 'quantum suicide'. Unfortunately, if the MWI is true (which is as yet unknown), then by committing quantum suicide you might prove this theory to yourself, but will never be able to prove it to anyone else[44].

The MWI is not the only theory that proposes the existence of other universes. In recent decades, there have appeared a number of theories in which our universe is only a unit in a potentially infinite number of universes. To mention just a few, hypotheses were proposed for a quilted universe, an inflatory multiverse, a brain multiverse, a cyclic multiverse, a landscape multiverse, a holographic multiverse and a simulated multiverse[45]. For example, it was suggested that the simulated multiverse exists on a complex system of computers that is capable of simulating entire universes. An interesting question is *who is the holder of this magical system of computers that can simulate not only stars and galaxies, but life and human consciousness as well?* And of course, none of these theoretical constructions can be verified in experiments.

Finally, 'string theory' replaced the point-like particles of matter, such as protons and electrons, with one-dimensional objects called 'strings'[46]. This theory's advantage is that it explains all known types of elementary particles through these particles' 'quantum states'; it also explains all four known fundamental forces – electromagnetism, the weak and strong nuclear forces and gravitation. A distinguished feature of string theory is that it postulates, on top of the known four dimensions (three in space and one in time), other dimensions as well. For example, in the 'superstring' version of string theory, there are 11 dimensions, with the extra seven dimensions being packed into a compact 'ball' called Calabi–Yau space. Despite all the advantages of string theory, it has been impossible thus far to test this theory in an experiment because of the unimaginably small size of the strings (the so-called Planck length) and the inaccessibility of the 'extra dimensions'[47]. An extension of string theory is 'M-theory', which tries to unite all of the versions of string theory. M-theory suggests that strings are one-dimensional versions of a yet more fundamental entity – the two-dimensional membrane, which vibrates in 11-dimensional space. Some physicists call M-theory 'a theory of everything', because in this theory all the diversity of the physical universe is reduced to one theoretical construct. *Because the vibration of strings and membranes is a final 'point of reference' and doesn't conform to any natural causes, this vibration must be a magical phenomenon.* This might be a reason as to why some physicists decipher the term 'M-theory' as 'magical' or 'mystical' theory[48].

As the idea that the subjective reality of an observer participates in the structure of the universe became increasingly accepted in physics, the degree of this participation grew. American physicist John Archibald Wheeler condensed this idea in his expression 'it from bit'. He writes, "It from bit symbolizes the idea that every item of the physical world has at bottom – at a very deep bottom, in most instances – an immaterial source and explanation; that what we call reality arises in the last analysis from the posing of yes–no questions and the registering of equipment-evoked responses; in short, that all things physical are information-theoretic in origin and this is a participatory universe"[49], p. 311. This means that scientists, through observing devices, ask nature questions that have only two possible answers: yes or no. When answered, a question like that provides a unit of information, or a 'bit'. Wheeler's approach was to put the unit of information at the foundation of quantum mechanics, which included the observer in the construction of reality. The extension of this theory was the 'information theoretic approach' – the attempt to derive a version of the quantum mechanical world from 'universal bits of information' on the basis of certain principles of general probability theory[50]. According to this theory, there is no need to postulate any 'objective' physical reality out there, and the Kantian 'thing-in-itself' could come to rest; instead, the regularities of the universe can emerge automatically from human subjective experiences and probabilistic algorithmic rules; because of these rules, random events condense into something that resembles the laws of nature as we know them from mainstream quantum physics.

Unfortunately, none of these theoretical constructions can be empirically tested. These hypotheses are about the 'final cause' of both the observed universe and the hypothetical other universes that cannot be observed currently or in principle. Because the final cause cannot be further reduced to any natural causes, it is by definition a 'something from nothing' type of magical event. The interesting twist is the aforementioned information theoretic approach to quantum mechanics, which entails that *random events, which are essentially magical, are capable of organising themselves into regular patterns that resemble the laws of nature.* Through the hypotheses that tacitly imply a magical event, the implicit belief in the supernatural, which thus far has been locked deep in scientists' subconscious, claims its right to public attention.

Thus far, we have been primarily concentrating on how magical phenomena infiltrate the physics of the micro world. What is happening in the physics of the mega world – cosmology?

Magic of the mega world

It has always seemed to me that the Big Bang is an event that is no less magical than is the biblical version of the world's creation by God in six days. In an infinitesimally small fraction of a second (and, as far as the human perception is concerned, instantly), the universe emerged from virtually 'nothing' – an incredibly dense and small entity called the initial singularity – and has been inflating ever since[51]. By definition, *when something as big as the universe emerges from something the size of an almost mathematical point, this is an 'emergence from nothing' magical phenomenon, which escapes the known laws of physics.* What makes this event look even more mysterious is that in this magically emerged universe there appears to be an observer who is able to understand this universe and is puzzled by the impossibility of logically explaining its origin.

Pondering the role of an observer in the universe, British physicist Jim Baggott recalls an old philosophy problem: if a tree falls in the forest and there's nobody around to hear it, does it make a sound? According to Baggot, the answer depends on how we define sound. There is no sound as a subjective quality, but there is sound as 'auditory waves in the air' [[13]]. This answer is insufficient, because the notion 'auditory waves in the air' is a result of human intellectual activity; hence, this notion (as with all notions of science for that matter) is a product of human consciousness. This implies that without a sentient being that hears the sound or thinks about the sound, the sound isn't there at all. Then what is there? There is 'something' about which we can say nothing except: if this 'something' interacts with our ears, we experience what is called 'sound', which we understand by creating the concept 'sound waves in the air'.

The same applies to the universe as a whole. Without a person who experiences the sounds and colours of the universe and creates theories about what they are, there is only 'something'. There are no photons and electrons, no black holes, no stars and no planets without a human being who thinks and reasons about these entities. One might ask: does this mean that people, like artists, create the universe as they wish? Of course it doesn't. The universe exists independently of our consciousness as 'something' out there. But *what this 'something' really is like we can only find out if we 'mix' this 'something' with our consciousness, like we mix sugar with water –* literally 'mix with' and not 'get through' our consciousness, like a light beam gets through clear glass. The beam of light gets through clear glass unchanged, but the 'something' of the universe, when it mixes with our consciousness, creates a unique 'fusion'. *This fusion of things-in-themselves and human mental reality we call electrons, stars and galaxies.*

It is *because the physical universe is a part of our own mind that we can understand the laws of nature.* Something that is not a part of our mental reality, like other universes, cannot be understood in principle. Exactly because the physical world is a part of our mind, we observe the wonderful 'tuning' of the universal physical constants to each other in such a way that their 'ensemble' makes it possible for life and humans to exist[52]. Indeed, if one of those fundamental constants (e.g., the gravitational constant g, which defines the speed of the universe's expansion) were just a little bit different, life in the universe would be impossible. Likewise, if the universe were a little bit younger or older than it is now, there would be no us, because there would be no necessary elements from which our bodies are built. The problem arises as to who tuned the 'piano of the universe' so finely? Who was that wizard that made the fundamental constants fit so perfectly with each other that life became possible in the universe?

The awareness of this problem gave rise to the 'anthropic principle' in cosmology, introduced by Australian physicist Brandon Carter in 1973[53]. The *weak anthropic principle* suggests that the universe's fine tuning is a random event, and because our universe is so nicely tuned for us to exist, we can only observe this universe and none of the others[54]. The weak anthropic principle leads to a conclusion that there must be an infinitely large number of universes from which our universe is the only lucky one to have human observers.

The *strong anthropic principle* goes much further than the weak anthropic principle and proposes that the existence of conscious observers (and therefore life) is not a random event, but a necessary condition for the existence of the universe. Conscious observers participate in shaping the universe as it is. This version of the anthropic principle is compatible with the Copenhagen interpretation, as well as with information-based versions of quantum theory, but it also allows for the possibility of the classic argument of intelligent design (see Chapter 1): the universe being designed so cleverly that it could not have emerged on its own. Because the strong anthropic principle does not exclude the Creator, many scientists argue in favour of the weak anthropic principle, which is compatible with the MWI of quantum mechanics (see Chapter 5 for more on the anthropic principle).

But the hope of avoiding the Creator by downgrading the anthropic principle is an illusion. As argued in the Introduction, a random event in its own right is a non-causal 'something from nothing' magical event. *Instead of the one purposefully created universe that the strong anthropic principle implies, the weak anthropic principle allows for the possibility of an infinite number of magical events.* An attempt to deal with this problem is proposed in the aforementioned information theoretic approach to quantum theory [[50]]. This approach aims to show that, if certain rules are applied to random events, these events have the potential to build themselves into regular patterns that are similar to those we call the laws of nature. But these rules or postulates (e.g., continuous reversibility, tomographic locality or existence of a fundamental information unit) are not arbitrarily selected; they are based on existing general probability theory, which in its own right is based on experience and observation. Thus, these postulates can be interpreted as the information theory versions of fundamental physical constants, which brings us back to the strong anthropic principle.

Both versions of the anthropic principle implicitly involve an element of the belief in the supernatural in still another way. Indeed, as mentioned above, according to astrology, there is a magical kinship, or sympathy, between people and planets. But the 'pre-established harmony' between the human mind and the laws of nature is a core feature of the anthropic principle as well. *Hence the link between sympathetic magical thinking and the anthropic principle*: astrologers can read a person's fate by using the 'language of planets', and modern scientists can 'speak' to nature using the language of physical laws.

Conclusion: from the Standard Model to magical physics

We started this chapter with the following question: how and why did it happen that some influential theories of modern physics and cosmology describe events that are indistinguishable from magic?

Searching for an answer, we turned to recent psychological studies on magical thinking and magical beliefs in modern people. These studies revealed that magical thinking is wired into the human mind and shows up in dreams, art and fantasies. Being confined to the realms of the imagination and the subconscious, magical thinking doesn't contradict modern peoples' belief in science, and peacefully coexists with logical thinking in peoples' minds. By contrast, the belief in magic in modern industrial cultures is viewed to be a fallacy and is thought to remain only in children and a small number of superstitious adults.

Yet psychological experiments of recent decades have shown that modern educated adults who consciously consider themselves non-believers in magic believe in the supernatural in their subconscious. The dominant belief in science prevents rational people from admitting their hidden belief in magic. Nevertheless, in certain circumstances, the implicit belief in the supernatural gains access to the realm of reflective thinking and rationally controlled behaviour. When this happens, rational adults openly acknowledge that magical phenomena are not just fantasies, but can produce effects in the real world.

The subconscious belief of rational people in the supernatural was one of the factors that contributed to the 'merger' between magical thinking and physical theories. This hidden belief created a psychological predisposition for magical phenomena to be invited into the realm of rational theorising. This predisposition alone, however, was far from being sufficient to allow the 'merger' to happen. Brought up by classical physics, modern specialists on quantum theory and cosmology are strongly opposed to the idea that their theories could benefit from magic in any way. Some other factors had to join the psychological predisposition in order for theorists, though reluctantly, to accept magical phenomena into their theoretical constructions.

The second factor that contributed to the 'merger' was *the link between magical and scientific types of thinking* in the search for truth. Magical thinking operates in the sub-conscious through symbols and is a 'generator of ideas'; it supplies scientific thinking with a pool of 'theories in the making'. Scientific thinking selects from that pool those theories that resonate with objective reality, and it rejects those that don't fit. In the process of this selection, scientific thinking uses two main criteria: verification by experimental facts and compliance of the 'raw theories' with the general body of existing physical knowledge. The link between magical and scientific thinking is a necessary condition of scientific discoveries. This link creates an 'umbilical cord' through which magical phenomena could penetrate into scientific theories. Usually, however, such penetration is prevented by experimental verification of the 'raw theories' supplied by magical thinking. Even combined with the hidden belief in magic, the link between magical and scientific thinking for centuries failed to advance magical phenomena into the world created by physical science.

The third – and crucial – factor that caused the aforementioned 'merger' is *the progress of physical science itself.* As physics proceeded from the macro world into the micro and mega worlds, the concept of a physical object changed. From an object that can be seen with the naked eye, the physical object turned into an object that must be deduced from the traces it leaves in various mediums. No longer supported by direct sensual experience, the physical object increasingly depends on the structure of observing devices. The incredible sensitivity of these devices made the fundamental link between human consciousness and physical objects evident. As a result, magical phenomena, such as direct 'Self over matter' and 'participation' phenomena, tacitly appeared in physics under the names of the Copenhagen interpretation and quantum entanglement, and in cosmology under the title of the anthropic principle. Nevertheless, up until the 1970s, a bunch of quantum theories called the Standard Model[55] coordinated themselves with experimental facts. The magical phenomena in physics were proven empirically, yet these phenomena were interpreted as weird but natural ones.

Finally, when theoretical physics moved further away from empirical reality and into the realm of such theories as string theory and the many-worlds theory, experiments became impossible. A major reality check criterion – measurements and experiments – needed to separate rational constructions concordant with observed phenomena from magical phenomena (see Figure 3.1) disappeared. The remaining criterion – compliance of the 'theories in the making' with the general body of current physical knowledge – is significantly more flexible and 'softer' than the experimental reality check. The thinning of the border between magical and scientific thinking, combined with the tacit belief in the supernatural, made it possible for magical phenomena to filter through into some theories of modern physics. What in the world of classical physics is magic (e.g., the creation of a universe by an act of consciousness) in some theories of modern physics (e.g., the 'many-worlds' theory, the 'information theoretic' theory) is admitted as reality. Physical scientists, who otherwise reject magic as superstition, become more benevolent when magical events come to them clad into sophisticated mathematics and are inaccessible to the experimental reality check.

Is the merger between magic and science a step forwards to the truth about reality or is it a step backwards to medieval astrology and alchemy? I think it's a bit of both, and that this merger was destined to happen. There was a time when physics separated itself from magic by dismissing the possibility of magical phenomena; this separation allowed humankind to make an explosive leap forwards in its capacity to control nature. What made this leap possible was the idea of objectivity of knowledge rooted in the invention of the scientific experiment. But, with physics rejecting the supernatural, the *primary role of subjective experience in creating scientific concepts* (see Chapter 3 for more on this) was only temporarily suspended.

Now, physical science has approached the border where again, at a more advanced level, it faces the phenomenon of magical participation – the deep level of unity between a person and the universe. Beyond this border lies the realm of concepts that are so general that experimenting with the underlying reality becomes impossible. Some physicists stop at this threshold, but others, armed with mathematics and imagination, venture into this uncharted territory.

And the belief in the supernatural, which thus far has been locked in the dungeon of the subconscious, is filtering into the very heart of physics – theories of the origin and structure of the universe.

Notes

1 Subbotsky, E. (2010b)
2 Kaku, M. (2008)
3 https://en.wikipedia.org/wiki/Fundamental_interaction
4 https://en.wikipedia.org/wiki/Neuroprosthetics
5 Lévy-Bruhl, L. (1985/1926)
6 www.almaclassics.com/excerpts/Ruslan_Lyudmila.pdf
7 Subbotsky, E. (2001)
8 Nemeroff, C. & Rozin, P. (2000)
9 Subbotsky, E. (2005)
10 http://ru.wikipedia.org/wiki/Принцип_относительности
11 http://newfiz.narod.ru/rel-opus.htm
12 https://arxiv.org/ftp/arxiv/papers/0711/0711.1145.pdf
13 Baggott, J. (2013)
14 Horgan, J. (1997)
15 Isaakson, W. (2007)
16 Stent, G. (1969)
17 Wilson, C. (1971)
18 Whitley, D. S. (2008)
19 Leroi-Gourhan, A. (1968)
20 Mithen, S. (2005)
21 http://ru.wikipedia.org/wiki/Алхимия
22 www.logosapologia.org/exo-vaticana-giordano-brunos-extraterrestrial-diabolicus/#_edn2
23 www.gumer.info/bibliotek_Buks/Psihol/freyd/ya_ono.php
24 Mamardashvili, M. & Pyatigorskii, A. (2009)
25 https://en.wikipedia.org/wiki/August_Kekulé; http://en.wikipedia.org/wiki/Ouroboros
26 http://en.wikipedia.org/wiki/Maxwell%27s_demon
27 Berne, J. & Radunsky, V. (2013)
28 http://harpers.org/blog/2009/02/einsteins-human-cosmos/
29 http://en.wikiquote.org/wiki/Albert_Einstein
30 https://en.wikipedia.org/wiki/Objectivity_(science)
31 Frazer, J.G. (1922)
32 Einstein, A., Podolsky, B. & Rosen, N. (1935)
33 http://en.wikipedia.org/wiki/Quantum_entanglement
34 Eberhard, P. H. & Ross, R. R. (1989)
35 https://phys.org/news/2017-07-physicists-retrocausal-quantum-theory-future.html
36 http://en.wikipedia.org/wiki/Copenhagen_interpretation
37 http://en.wikipedia.org/wiki/Double-slit_experiment
38 Donati, O., Missiroli, G. F. & Pozzi, G. (1973)
39 http://laser.physics.sunysb.edu/~amarch/eraser/

40 http://laser.physics.sunysb.edu/~amarch/eraser/Walborn.pdf
41 Everett, H. (1957)
42 DeWitt, B. S., Graham, R. N. (Eds) (1973)
43 Green, B. (2011)
44 Tegmark, M. (1998)
45 https://en.wikipedia.org/wiki/Multiverse
46 http://en.wikipedia.org/wiki/String_theory
47 Woit, P. (2007)
48 http://en.wikipedia.org/wiki/M-theory
49 Wheeler, J. A. (1990)
50 Koberinski, A. & Müller, M. P. (2017)
51 http://en.wikipedia.org/wiki/Big_Bang
52 Davies, P. (2006)
53 http://en.wikipedia.org/wiki/Anthropic_principle
54 Barrow, J. & Tipler, F. (1986)
55 https://en.wikipedia.org/wiki/Standard_Model

5

THE PULL OF ETERNITY

Hope for immortality as a belief in the supernatural

Problem: why are we not solipsists?

In his autobiography, British film director Michael Winner mentioned that his earliest memory was a lamplighter igniting the gas jets on lampposts. "One day," he wrote, "the lights will go out… Movie lights will shine on. But not for me. The lights gripped me from the beginning and they will until the end. I've spent my life among them. They've lightened the dark"[1], p. 2. This image of the unavoidable cycle of a person's life reminded me of the simple truth known to the ancients: man is the measure of all things. A human person is that very 'god' who creates and supports his or her private universe. The person will pass away and his or her private universe will cease to exist, as silently and completely, as it appeared when the person was born. There will be no more physical theories, megaparsecs of space and billions of years of time. There will be no need to be concerned about the fact that sometime in the distant future the sun will swallow the earth and there will be no place for people to live. Surely, our private universe will 'burn out' much sooner than that. What will happen after that is of no concern. Philosophers call this point of view 'solipsism'.

"But our children and our deeds will remain," a reader might say. That is true. But for how long? What time is left for humanity to exist? According to most anthropologists, modern humans have been around for approximately 200,000 years, civilisation 12,000 years and human history 4,500 years[2]. But the universe has existed for 14 billion years and the earth for 4.5 billion years. Cosmologists predict that in 1.1 billion years, due to increased solar activity, the oceans on earth will evaporate, and in 7.5 billion years our sun will turn into a red giant and annihilate earth[3]. According to the cosmic calendar, if the age of the universe takes a year and the age of the solar system takes around four months, then modern humans have existed for seven minutes, history from the beginning of literacy has lasted for nine seconds and the longest individual human life lasts for 0.25 seconds[4]. It is clear that the existence of modern humans takes up very little time compared to the existence of the universe, and the length of human history proportionally is still much smaller. Humankind lives on the scale of thousands of years, and the universe on the scale of billions of years, which is a million times longer. There are a number of scenarios that could bring humankind to its death and do so fast, including nuclear annihilation, biological warfare, a global pandemic, ecological collapse, global warming, a meteor impact and volcanism[5]. In other words, on the cosmic scale, the life of humankind, not to mention that of a human individual, is incomparably short and unimaginably fragile, and if you are not a fiction writer then searching for knowledge on the distant future of the cosmos beyond a few thousand years makes no practical sense.

And yet we observe an amazing phenomenon of 'cosmic thinking': billions of dollars are spent every year on studying galaxies that are on the edge of the visible universe and on discovering quantum particles that can only have theoretical significance. For instance, the cost of the construction of the Hubble Space Telescope is estimated at over $2.5 billion, the total operating budget of the Large Hadron Collider (LHC) runs to about $1 billion per year, while the total cost of finding the Higgs boson with the help of the LHC ran to about $13.25 billion. Renowned physicists and cosmologists consider various scenarios of the end of the world (e.g., the heat death of the universe or the collapse into a singularity point similar to the one that gave birth to Big Bang)[6] and the end of the earth [3], specialists on cybernetics discuss the possibility of replacing humans with intelligent machines[7] and turning the whole universe into an omnipotent and omniscient computer[8] and philosophers contemplate the attainment of the state of divine unification by our universe in the hypothetical 'Omega Point'[9]. Ordinary folk like ourselves also can't imagine the world without us. Thinking of the past or the future, we are always present there in person, as if looking at this imagined world 'from above'. We can even think about existence prior to the Big Bang, without the universe, without space and time, but amazingly we are still there. In other words, consciously or unconsciously, we live our lives as though we were immortal. In people who believe in God, the hope for immortality of the soul (IS) is assigned by their faith, but *why is this hope implicitly present in people who consciously deny their belief in God and in the supernatural? In other words, why are we not solipsists?*

In this chapter, we will discuss the psychological grounds of the phenomenon of cosmic thinking. We will try to explore the possibility that at the base of this phenomenon is rational adults' implicit hopes for IS and the immortality of humankind. Although subtle differences exist between the concepts of 'the soul' and 'the mind', in this chapter these concepts will be used interchangeably. We will try to investigate whether recent studies on magical beliefs in modern people could provide new insights into the causes of the phenomenon of cosmic thinking.

Belief in the supernatural and the hope for immortality

For Babylonians, Sumerians and early Egyptians, the planets, the sun and the moon were gods – living entities capable of hearing and understanding human prayers[10]. These gods ruled the cycles of the year, rainfalls and sea tides. Even the founder of European philosophy, Thales of Miletus, who lived in the fifth century BCE, believed that all things were full of gods[11]. As argued in Part I of this book, the belief in gods and spirits comes as a result of the belief in IS. Ancient burial sites with human artefacts indicate that people started to believe that physical death was not the end of a person, and that a person's soul survived physical death and proceeded into a special invisible realm, where it became a spirit. The spirits of dead ancestors became worshiped as the first gods. Eventually, the belief in spirits was extended from people to animals and natural things, giving birth to the belief in animism. If animals and even physical objects have spirits inside them, then it is possible for a human person to address those spirits and affect physical objects spiritually via magical spells and rituals. This belief in the ability to affect nature by the sheer effort of the mind became known as magic, or the supernatural (see the Introduction). From the above, it follows that *the belief in magic is linked to the belief in the IS, and vice versa*. And indeed, as history shows, the belief in magic and in IS go hand in hand, culminating in mummification rituals of early Egyptians, Greek myths of the underworld and Christian belief in the afterlife.

To investigate whether the belief in magic and IS are linked one to the other in modern people, an experiment was designed. Participants (university graduates and undergraduates in the Soviet Union and West Germany) were shown an apparently magical phenomenon: an experimenter transformed a physical object by sheer power of his mind. The participants were then interviewed on whether they believed that they had witnessed an instance of real magic, and also on whether they believed in IS and in the existence of the Supreme Being. The results revealed a significant correlation between the beliefs in magic and IS; the belief in IS also significantly correlated with the belief in the Supreme Being[12]. This shows that modern adults who openly acknowledge their belief in the supernatural (e.g., direct Self over matter magic) also acknowledge their belief in IS. But what about those rational adults who consciously deny their belief in the supernatural? Why would modern physicists and cosmologists believe in IS, if most of them consider themselves non-believers in magic or in God?

The answer could be found in recent psychological studies on magical beliefs in modern people. These studies have shown that 'at the bottom of the mind', in the realm of subconscious, most rational adults continue to believe in magic. An example of such belief is the belief in the law of sympathy – the belief that there is a supernatural link between objects that represent one another, while having no physical, causal link one to the other at the same time (see Table 4.1 for more on the laws of magic). For example, in medieval Britain, people protected themselves from witchcraft by making a doll of the witch out of rags and then piercing it with pins, with the intention of causing physical harm to the witch and breaking their charm[13]. With the aim of examining whether the belief in magical sympathy exists in educated people today, participants (university students) were asked to stick a needle into a doll. The doll represented either a person who, by his or her improper behaviour earlier in the experiment, made the participants think badly of him or her or a person who had displayed a positive pattern of behaviour. When later the person who the doll represented started complaining about having a bad headache, participants who had been made to have negative feelings about the person acknowledged that they felt responsible for the person's misfortune, as if their manipulation of the doll was the cause of the person's headache[14]. Other studies have revealed that, under certain circumstances, most modern rational adults openly acknowledge that they believe in direct Self over matter magic[15,16].

The link between the belief in magic and in IS, supported by historical analysis and psychological studies, suggests that *the subconscious belief of modern rational adults in the supernatural must be accompanied by an implicit hope for IS.* This might provide an answer to the main question asked in this chapter: why modern rational adults, who openly deny IS, behave as if the human mind and humankind are immortal.

Having taken on board the hypothesis that the belief in the supernatural implies the hope for IS, let us look more closely at how it happened that the belief in IS, which came to the ancients naturally, was abolished in the modern world, and how modern people react to this fact.

A person and the universe: conflict of the worlds

The ancients believed that they were at the centre of the world. They lived on the earth. Around the earth, gods–planets revolved, with which humans communicated through prayers and rituals and into whose realm human souls passed after death. People believed in the cosmic link between a person and the surrounding world.

Gradually, however, the ancient's vision of the gods changed: the behaviour of the gods became less capricious and more predictable. The Antikythera mechanism – an ancient analogue computer created around 100–150 BCE and used to predict astronomical positions and eclipses for calendrical and astrological purposes – symbolises this shift in the ancient mind. Such 'restraining' of the cosmic gods' behaviour later spread to the whole of nature and gave birth to the notion of 'the laws of nature', and eventually to modern science. Due to the achievements of science in the seventeenth to nineteenth centuries, in industrial countries the belief in animism became the privilege of small children and a limited number of superstitious adults. As science rejected the belief in magic and animism, so did it reject the belief that a human soul (the mind) is immortal and survives the death of the body.

By stating that the laws of nature are independent of human observers, science erected a wall between a person and the universe. People started to feel themselves to live on a tiny island lost in the boundless ocean of the cold and indifferent universe. Even here on earth, in people's own homes, science did not leave people much ground for being proud of themselves, reducing all of their inspiration and achievements to the interplay of soulless, selfish genes[17]. Modern science contends that the soul is an illusion that accompanies the work of the brain, and when the brain stops functioning, the person vanishes from the world[18]. As British philosopher Bertrand Russell eloquently put it, "That man is the product of causes which had no prevision of the end they were achieving; that his origin, his growth, his hopes and fears, his loves and his beliefs, are but the outcome of accidental collocations of atoms … that all the labours of the ages, all the devotion, all the inspiration, all the noonday brightness of human genius, are destined to extinction in the vast death of the solar system, and that the whole temple of Man's achievement must inevitably be buried beneath the debris of a universe in ruins"[19].

Not surprisingly, in this soulless and heartless world, people started to feel uncomfortable. While in the magical universe gods gave meaning to the existence of both people and the world, in the scientific universe this meaning was lost. A conflict emerged between the human mind, which is desperate for meaning and eternal life, and the meaningless world created by science. And this conflict led to people trying to break away from the claws of the world that science had created. The belief in IS that still lurks in modern people's subconscious began to feed the hope for a miracle – for the chance that the scientific view of the world, with all its formidable, convincing power, was missing something important. A search for evidence that there are phenomena in the world that surpass the laws of nature began.

A search for evidence

One of these phenomena is the magical participation *between an individual's fate and the planets*[20]. According to the Alexandrian astrological tradition[21], a person's fate can be predicted on the basis of astrological charts. French mathematician Michel Gauquelin[22,23] statistically examined the 'Mars effect' – the astrologers' claim that people born just after the planet Mars raises or culminates become outstanding athletes in various sports. To his own surprise, he indeed found a significant statistical correlation between the fact of being born 'under the influence of Mars' and athletic talents. Later, similar correlations were established between Jupiter and one's acting ability and Saturn and one's scientific ability. British psychologist Hans Eysenk (known for his concepts of the personality dimensions 'introversion' and 'neuroticism'), who once viewed astrology with scepticism, checked the correlations and confirmed their validity[24]. Eysenk's colleagues Mayo and White[25] further extended the analogy and showed that people born under odd-numbered zodiac signs are usually extraverts and those born under even-numbered zodiac signs tend to be introverts. Later attempts to replicate Gauquelin's study failed and the study's effects were explained by selection bias and statistical inaccuracies[26]. Surprisingly, for myself the study worked well. Having read Gauquelin's results, I immediately checked my data (I was born on January 15); and indeed, all was in place: born under an even-numbered zodiac sign – Capricorn (I am a clear introvert) – and under the influence of Saturn (I devoted myself to a scientific profession).

Another form of magical phenomena are *paranormal psychological effects*, such as telekinesis – the direct effect of thinking on inanimate matter. Proving that a direct action of our Self could affect physical processes would radically change our view of the world. This is because this fact would show that something immaterial – a human subjective experience – is able to affect matter, and would do so not through one of the four known types of physical forces.

"Hold on," a reader might say, "aren't modern prosthetic devices reacting to brain signals an application of telekinesis?" No, they are not. As argued in Chapters 3 and 4, a prosthetic device doesn't react directly to a human thought; rather, it reacts to the patterns of electric signals that accompany the thought. It is the interaction between a person's Self and the electric activity of his or her brain that is indeed a magical phenomenon, whereas the brain connects to a prosthetic device strictly through a physical force (see Table 0.3).

The effects of our thinking on the electric impulses in the brain, as well as on our voluntary movements, are so mundane that they are not usually viewed as supernatural. In contrast, if the Self could directly affect matter beyond our body, this would be a persuasive proof that subjective experience is not an illusion, but has a real 'presence' in the physical universe. Interestingly, the possibility of this kind of direct influence of a human thought on certain physical processes has been examined experimentally, and the studies indeed found statistically significant effects. Although such effects are weak and cannot be used for any practical purposes, they are nevertheless real[27,28,29]. Being curious and a disbeliever in parapsychology, I attempted two studies myself: one on telekinetic effects[30] and another (together with extrasensory perception [ESP] researcher Adrian Ryan) on ESP[31]. To my surprise, despite some inconsistencies in the results across experiments, statistically significant positive paranormal effects were obtained in both studies (see Chapter 6 for more on these studies). Nevertheless, today most scientists believe that paranormal effects lack sufficient experimental support and are impossible to replicate.

One more manifestation of the supernatural that directly leads to the hope for immortality are '*near-death experiences*'. In his book *Life after Life*, American psychologist and medical doctor Raymond Moody reviewed hundreds of cases in which people who had been pronounced clinically dead but were subsequently resuscitated described their memories[32]. These memories include such episodes as leaving the body, levitation, an overwhelming feeling of peace, security and warmth, a feeling of unity with the universe and fast movement along a tunnel towards a source of light. The important element of this extraordinary experience is obtaining a new view of themselves and the universe. According to a Gallup poll, in the USA alone, around eight million people have reported this kind of experience[33]. Interestingly, in a considerable number of such patients, the fear of death disappears. Yet, at present, most scientists interpret near-death experiences not as proof of life after death, but as hallucinations elicited by the dying brain.

Still another manifestation of our implicit faith in the magical unity between a person and the universe is *science fiction*, served with existing facts and spiced with enthralling speculations. Books by Erich von Däniken gave rise to the Ancient Astronauts theory[34]. According to advocates of this theory, halos over the heads of human figures in prehistoric rock paintings represent the helmets of the astronauts' spacesuits, and the miracles described in myths and in the Bible are the observers' naïve reports of the aliens' technological wonders. These ancient astronauts performed feats that people of the past could associate only with the gods: create human–animal hybrids via genetic engineering, move huge stone boulders by cancelling gravity, melt stone with a laser beam and fly in the air and space. At that, the proponents of this esoteric theory claim that magic has nothing to do with these wonders and all of the 'miracles' were the achievements of science that was thousands of years ahead of modern human science. If the omniscient, benevolent aliens visited earth in the past, then there is hope that they might come back in the future and share with us their knowledge on how to live longer lives and travel to other galaxies and universes.

Those scientists who feel uncomfortable about parapsychology and science fiction and yet reject the idea of God put their faith in *humanism* – the belief in the unlimited power of human reason. Humans will save themselves through the unstoppable progress of reason, science and technology. But overcoming all the dangers that await humankind in the future implies a belief in humans' extraordinary powers, thus putting humans in the place of gods. Soviet physicist Nikolai Kardashev introduced the hypothetical classification of civilisations according to the type of energy consumed. Type I civilisations use the energy of their planets, type II civilisations can utilise the energy of their stars and type III civilisations can employ the energy of their galaxies[35]. By analogy, Michio Kaku classified technological achievements by their degree of impossibility to modern technologies[36]. For instance, he predicts that in a few thousand years it will be possible to overcome the 'impossibilities of type II' by creating spaceships that fly faster than the speed of light, travelling back in time and travelling to other worlds through the 'wormholes' that connect different universes. Sir Martin Rees, the UK Astronomer Royal, writes: "Wormholes, extra dimensions, and quantum computers open up speculative scenarios that could transform our entire universe eventually into a 'living cosmos'!" (cited in [36], p. 281). French philosopher and anthropologist Pierre Teilhard de Chardin brought a philosophical basis to the concept of humanism by directly pointing out the magical participation between humankind and the cosmos: "In its present state, the world would be unintelligible and the presence in it of reflection would be incomprehensible, unless we supposed there to be a secret complicity between the infinite and the infinitesimal to warm, nourish and sustain to the very end – by dint of chance, contingencies and the exercise of free choice – the consciousness that has emerged between the two. It is upon this complicity that we must depend. Man is irreplaceable. Therefore, however improbable it might seem, he must reach the goal, not necessarily, doubtless, but infallibly" [9], p. 275. But science fiction and humanism, with all the thrill and optimism of their futuristic prognoses, provide no evidence to support the hope for IS.

To summarise, exploration of extraordinary phenomena has thus far failed to find convincing evidence that the hope for IS really does have a ground. Surprisingly, some evidence came from an unexpected source – modern physics.

The anthropic principle

According to classical physics, the universe exists independently of humankind. However, discoveries in quantum physics put this view under question. In order to understand how this happened, we need to look more closely at what exactly it means 'to exist'.

Let us imagine a planet that circles the sun somewhere at the far end of the solar system and thus far is totally unknown to humans. Does this planet exist or doesn't it? We cannot answer this question until our devices register this planet or its impacts on the planets that we already know of. For instance, the existence of the planet Neptune had been predicted on the grounds of deviations of the known planet Uranus from its orbit. On the same grounds, in 1906, it was predicted that there was still another planet that was influencing Uranus' orbit; the unknown planet (called 'Planet X') was registered only in 1930 and named Pluto[37].

So, what does it mean, 'to exist'? In regard to a human person, it means two things. First, for a person to exist in the proper sense of this word, the person has to be aware of himself or herself. When a person is under general anaesthesia, the status of the person's existence downgrades to a lower level – to their existence in the minds of others. Second, for a person to exist, the person has to have memories of his or her past. A person who lost memories of his or her past is a different person; for this person, the person who existed prior to the complete loss of memory is dead. The person with no autobiographical memories can restore such memories only by what other people tell him or her, thus relying on beliefs rather than on knowledge.

And what does 'to exist' mean in regard to a planet? Modern science teaches us that planets aren't aware of themselves and don't have autobiographical memories. Even if we tell the planet its past history, it won't be able to understand the story. It turns out that the planet exists only if people know about its existence. Without people's knowledge about the planet, the planet doesn't really exist. Philosophers like David Hume[38], George Berkeley[39] and Arthur Schopenhauer[40] pointed out long ago that the world and the human representation of the world are tightly connected. Modern psychology confirms that a human being can interact with the world solely through his or her subjective experiences – sensations and perceptions. Our conscious representations of the world are built on the ground of our sensations and perceptions as well[41]. Contrary to common sense, it appears that without humankind, the universe doesn't really exist.

For a long time, physical science resisted this view. Indeed, for instance, don't the fossils that we find in the ground tell us that tens of millions of years ago dinosaurs roamed the earth, and don't meteorites that fall from space reveal that before humans emerged on earth our planet was already around 4.5 billion years old? The problem is however that humankind, like the person who lost his or her autobiographical memory, learns about the past from the facts it observes today. Because we could not observe the universe before humankind came into existence, we have no choice but to believe what the fossils, meteorites and the light coming from the stars tell us about our past and the past of our universe. But the same facts can be interpreted in a variety of ways. For example, advocates of 'scientific creationism' interpret many facts of geology and palaeontology to provide scientific support for the creation myth described in the Book of Genesis[42]. Perhaps the creationists' interpretation of the facts is not as logical and consistent as the one provided by the theorists of evolution, but it's the difference in the views that matters. The difference between creationists' and evolutionists' views on the same facts shows that we know our past by interpreting the facts we observe in our present. Even in the natural sciences, interpretations of known facts can change, which alters our theories about the past. For example, quite recently the facts that fossils of the same species of animals and plants were found on different continents separated by oceans were explained by the existence of hypothetical land bridges which would now have become submerged. It was only when the theory of plate tectonics was developed by Samuel Carey in 1958 that the idea of continental drift (originally put forward by Abraham Ortelius in 1596 and more fully developed by Alfred Wegener in 1912) was finally accepted[43]. As a result, the theory of land bridges was dropped and replaced with the theory that all continents separated from one giant supercontinent, Pangaea. Again, contrary to common sense, it looks as though we create the past of our planet and the universe by explaining the facts we observe in the present.

The realisation of the inseparable link between our knowledge and the way we interpret observable facts gave birth to the 'anthropic principle' (see Chapter 4). Introduced by Brandon Carter in 1973 at the Krakow symposium to commemorate Copernicus' 500th birthday, the anthropic principle was a reaction to the 'Copernican principle'[44]. Whereas the Copernican principle states that humans have no privileged position in the universe, Carter proposed that, to some extent, humanity's position in the universe is inevitably privileged[45]. There are two versions of the anthropic principle. According to the weak anthropic principle (WAP), the fundamental constants of our universe (e.g., the gravitational constant, the mass of a proton, the age of the universe) are such that they make the existence of intelligent observers possible. For example, if our universe were a little younger or older than it is now, life could not emerge because there would be no necessary elements from which living bodies are built. Exactly why in our universe the fundamental constants are just right for making intelligent life possible is usually explained by selection bias, which means that this perfect fit is a random event. But this explanation comes at a price: it means that although our universe is associated with the existence of intelligent observers, this association is only coincidental and partial. For a selection bias to happen, there has to be a pool of variations; on this ground, the WAP has to accept that our universe is one of many universes, or a part of the multiverse. But most importantly, it doesn't follow from the WAP that the universe didn't exist prior to the emergence of intelligent life and will cease to exist with the end of intelligent life.

In contrast to the WAP, the strong anthropic principle (SAP) holds that life is not just a consequence of a lucky combination of fundamental physical constants, but that intelligent observers participate in building the picture of the universe[46]. As discussed in the previous chapter, according to the 'Copenhagen interpretation' of quantum mechanics, every elementary unit of matter, such as a photon, does not have definite physical properties until it is registered by observation[47]. This means that the observer's mind doesn't simply reflect reality like a mirror reflects a person, but *participates* in the way the reality 'condenses' into something definite. In his participatory principle, American theoretical physicist John Wheeler rephrased this thought as 'it from bit'[48]. As explained earlier in the book, this expression suggests that the way a quantum event reveals itself to us depends on the way the observer, via observing devices, asks nature questions, which can only have two answers – 'yes' or 'no'. An answer to this binary question is a commonly accepted unit of information, or a 'bit'. It is important to note that, in Wheeler's view, the act of observation does not create reality; rather, the picture of reality depends on the form in which the observers ask the question: one way of asking the question 'provokes' electrons or photons to reveal themselves as particles, and another way makes the same elementary units of matter 'react' as waves (see Chapter 4 for more on this). Unlike the WAP, the SAP implies that a human person and the universe are locked in participation; one cannot exist without the other.

To summarise, it follows from the SAP that at the level of quantum events, supernatural phenomena, such as direct 'Self over matter' and participation, are indeed possible. This shows that the belief in the supernatural that most modern people subconsciously hold is not completely ungrounded. However, the SAP doesn't imply that the observers' consciousness is immortal. The SAP links intelligent observers to the physical universe for as long as both of them exist. Perhaps the universe is indifferent to its own existence and can perish, taking the observers with it into oblivion. Or maybe, and more likely, the observers die and the universe as we know it vanishes with them.

Even so, the hope for IS can survive by clinging to the idea that the individual consciousness does not have to vanish with the death of the physical universe, but can join some kind of 'cosmic consciousness' beyond the physical universe. But this hope runs into the problem of memory. As we know, a person can die before his or her physical death if they lose his or her autobiographical memory. Because memory is stored in the form of engrams (i.e., patterns of change in neural tissue in the brain), a halt in the brain functioning also means a complete loss of memory[49]. This means that even if the individual consciousness magically remains after death, this will be a consciousness without memory, and this 'memory-free' subject would never know who he or she had been prior to the moment of death. This problem can be overcome only if one assumes that apart from being rooted in the brain's engrams, an individual's memory also has a base in some other 'storage facility', which is situated beyond the human body, and even beyond the physical universe. Despite the extraordinary nature of this idea, there are theories that propose the existence of such 'out-of-body' carriers of an individual's autobiographical memory.

Memory beyond the brain

In Plato's dialogue *Meno*[50], Socrates' opponent Meno points out the 'paradox of knowledge'[51]. In essence, the paradox states that in order to enquire into anything, you have to already know the properties of what you are looking for, otherwise you won't be able to recognise the thing you are looking for even if you come across it. But if you know the properties of what you are looking for, then enquiry is unnecessary. On the other hand, if you don't know the properties of what you are looking for, then enquiry is pointless. Therefore, enquiry is either unnecessary or pointless. Looking for a way out of this paradox, Socrates puts forward the theory of 'anamnesis' – knowledge as recollection. He assumes that the human soul is immortal and goes through a series of reincarnations. This means that when the soul transmigrates into a new body, it already contains knowledge about the world; however, because of the trauma of birth, the soul forgets its knowledge. It therefore turns out that what we usually view as learning is in reality remembering of the already known. A teacher is not the holder of knowledge, but a person who helps a student to remember the knowledge the student knew but forgot. Socrates illustrated his theory by putting questions about a geometrical theorem to an uneducated slave boy. In the beginning, it looks as though the boy doesn't know the theorem, but gradually, by answering Socrates' questions, the boy comes to the correct answer. This demonstrates, Socrates says, that the boy knew the theorem, but was unable to remember it without leading questions. So, according to Socrates and his follower Plato, knowledge in the form of 'universal ideas' is stored not in a person's body, but in a certain depository beyond the body; this theory also implies that when a person dies, his or her knowledge doesn't disappear from the world, but is stored in this magical depository, together with the person's soul.

Of course, it might require some effort to start believing in the anamnesis theory. Nevertheless, in the second half of the twentieth century, there appeared experimental evidence that seemed to confirm this theory. Research has shown that young infants and even new-born babies possess knowledge that they could not possibly acquire through learning. To raise just a few examples, new-borns showed the ability to distinguish between canonical geometrical shapes (such as a cross and a circle)[52]; one-month-old infants could distinguish elastic objects from rigid ones and even transfer this understanding from the tactile modality into visual one[53]; at the age of four months, babies were able to infer that one solid body cannot go through another solid body unimpeded[54]; and 15-month-old children could predict errors in another person's behaviour if that person's actions were based on his or her false beliefs regarding where a toy had been hidden[55]. How could babies possibly show these cognitive skills if they hadn't had an opportunity to learn them through experience?

One might assume that these cognitive skills are innate and transmitted through genes. But this interpretation is unlikely, because genes can carry only a very limited amount of information. Scientists' early hopes that the genetic code could exhaustively explain the forming of an individual organism did not come true. As British specialist on molecular genetics Rupert Sheldrake writes, genes code linear sequences of amino acids in proteins, which still have to fold into complex three-dimensional forms. "Even if the protein-folding problem could be solved," he wrote, "the next stage would be to attempt to predict the structures of cells on the basis of the interactions of hundreds of millions of proteins and other molecules, unleashing a vast combinatorial explosion, with more possible arrangements than all the atoms in the universe"[56], p. 173.

In order to explain the process of the formation of cells, tissues and organs, Sheldrake introduced the concept of the 'morphic field' – a special, non-physical formation that 'tells' the cell to which organ of the body (e.g., a hand or a leg) the cell belongs and what functions it performs. The author illustrates the relationships between the morphic field and genes by the relationships between a TV programme we are watching and the technical structure of a TV set. The image on the TV screen is impossible to have without a complex technical scheme of transistors and electronic circuits that compose the TV's body, but the image is not created by this technical body. The image is created by electromagnetic waves that are captured by the TV's body and transformed into the image on the screen. In biology, the morphic field plays the role of electromagnetic waves – it contains the 'programme' of an organism – whereas genes (the parallel of the TV set) convert this programme into a whole organism. If a transistor is broken in the TV set, this may cause a defect in the image on the screen. In a similar way, a defect in a genetic scheme (e.g., a mutation) can cause a defect in the structure and behaviour of the organism. But this doesn't mean that the organism is directly determined by the combination of genes. Genetic mutations change the 'tuning of the aerial' of the genes, which start to 'catch' a wrong programme from the database contained in the morphic field. For instance, a fruit fly (*Drosophila melanogaster*) may grow a second (and useless for its ability to fly) pair of wings. Although Sheldrake doesn't associate the concept of the morphic field with the supernatural, such an interpretation is possible, considering that in his opinion morphic fields cannot be "reduced to standard chemistry and physics" [ibid., p. 173].

Sheldrake assumes that nature possesses 'inner creativity' and 'inner memory'. He maintains that the laws of nature are not given from the outset, but develop gradually like habits in humans and animals. For instance, new chemical substances develop spontaneously, like the known laws of physics developed soon after the Big Bang, but then these new substances begin to facilitate the emergence of the same substances at a distance through a mechanism Sheldrake calls 'morphic resonance'. If on some planets in the universe a certain chemical compound already existed, then the synthesis of this compound on earth would be faster and easier to achieve than the synthesis of a completely new compound that had never before been present in the universe. Sheldrake highlights a variety of facts in support of his hypothesis.

For example, the crystallisation of new chemical substances, once achieved in one particular laboratory, makes similar crystallisation in an independent laboratory on another part of the globe more easily achievable, without any 'know-how' exchange between these laboratories. From the perspective presented in this book, morphic resonance is a case of magical sympathy: the influence of an event on a similar event at an arbitrary distance, both in time and in space, and not through any of the known fundamental forces of nature.

But if nature has its own memory, then it should be able not only to remember, but also to forget. In humans, skills not only can be acquired, but can also die out (e.g., if a piano player or a sportsman doesn't practice for a long time, his or her quality of performance goes down). Similarly, one can expect that some of the regularities in nature may diminish and even disappear. Confirmation of this hypothesis can be seen in the unexplainable gradual decrease of the reproducibility of experimental phenomena in psychology and other sciences, known as the 'decline effect'. In the 1930s, American parapsychologist Joseph Banks Rhine found that in some of his participants their abilities to correctly guess other peoples' thoughts faded until they disappeared completely[57]. Later, a similar decline effect was observed in other branches of psychology. In 1990, American psychologist Jonathan Schooler reported the 'verbal overshadowing' phenomenon. According to this phenomenon, under some situations, putting non-verbalisable thoughts into words could be disruptive: participants' recognised objects they had seen and named worse than objects they had seen but didn't name[58]. However, over the years, Schooler found it increasingly difficult to replicate this effect. Looking for an explanation, he writes: "Perhaps, just as the act of observation has been suggested to affect quantum measurements, scientific observation could subtly change some scientific effects. Although the laws of reality are usually understood to be immutable, some physicists ... have observed that this should be considered an assumption, not a foregone conclusion"[59], p. 437. Schooler sometimes called the decline effect 'cosmic habituation'[60]. When we conduct a scientific experiment, we ask nature a question and expect a standard answer every time we repeat this question under the same experimental conditions. In science, this repeated emergence of standard answers is called 'replicability'. Yet it turns out that however hard scientists try to maintain exactly the same experimental conditions, nature's answers eventually fade. It's as if nature 'habituates' to stimuli that are always there. In living organisms, habituation is when an organism gradually stops reacting to a stimulus that happens repeatedly. To put it in other words, nature reacts like a living creature – it gradually 'forgets' its own answer. Later, a similar decline effect was found in other sciences, such as biology and pharmaceutics. For instance, drugs that used to provide a high therapeutic effect lost their therapeutic power with time, even though the patients who took them had never used them before [ibid.].

So, there are some grounds to assume that both the autobiographical memory of a human individual and the 'memory of the universe' (e.g., the laws of nature) can be stored in a certain hypothetical depository that is beyond the physical universe. Throughout history, scientists gave different names to this depository, such as '*the realm of ideas*' or '*the morphic field*'. The 'depository hypothesis' doesn't replace the traditional explanation of human memory by synaptic engrams, but rather complements this explanation. If this hypothesis is taken seriously, then the implicit hope for the immortality of an individual human consciousness and of human-kind for that matter, is not without some grounds after all. But still, it remains only a hope.

A person in the ocean of time: conclusions

We started this chapter with the question of the psychological causes of the phe-nomenon of 'cosmic thinking' when renowned scientists and even institutions engage in projects that seem to have no practical value to modern life. These projects include the exploration of problems such as the origin of the universe, the future of earth and the universe billions of years from now, the structure of objects that are situated hundreds of millions of light years away from the Milky Way or are so fundamental that can exist only in theory and the fate of humankind in the future measured on a cosmic scale. These studies require serious investments. But human civilisation, which has existed for only a few thousand years, is facing a number of urgent challenges (such as ongoing regional wars, refugee crises, global warming, undernourishment, lethal diseases, overpopulation and ecological disasters) and may come to an abrupt end at any moment due to a variety of lethal effects. We hypothesised that the acute interest in global cosmic issues has its basis in the sub-conscious hope of modern people that a person's mind and humankind itself are immortal.

The historical analysis revealed that the hope for IS is rooted in the beliefs of the ancients in the supernatural realm where gods and spirits lived and where souls passed after people's deaths. Modern science proclaimed that belief in the supernat-ural is a fallacy, and this imposed a ban on the hope for personal immortality.

Nevertheless, psychological research in recent decades has shown that the belief in the supernatural is alive not only in small children and superstitious adults, but in the majority of educated adults who consciously view themselves as non-believers in magic or in God. The research also supported the notion that the belief in magic is accompanied by the belief in IS. These results suggested a possible explanation for the phenomenon of cosmic thinking: in their subconscious, modern people believe in the supernatural, and this belief gives a person the subconscious hope that the individual human consciousness, and humankind, could be immortal.

We reviewed various attempts people have made to provide their implicit belief in the supernatural with evidence: studies of paranormal phenomena, exploration of the possibility of participation between a person and the universe (e.g., the 'effect of Mars', the direct effect of thoughts on matter, near-dearth experiences), cre-ating quasi-scientific theories (e.g., the 'ancient astronauts' theory) and developing humanistic approaches (e.g., the 'Omega Point' theory). We came to the conclu-sion that none of these attempts provided solid evidence for supernatural phe-nomena, which leaves the implicit hope for personal immortality without support.

In contrast to the aforementioned attempts to prove the existence of the supernatural, the achievements of quantum physics showed that at the level of quantum events the magical participation between human consciousness and the physical universe is indeed a reality. These achievements were summarised in the SAP. But the SAP does not guarantee the immortality of humankind, as it links the physical universe to intelligent observers for only as long as the observers live.

Finally, we examined the assumption that the mind can survive the death of the physical universe by joining the cosmic consciousness. We found that despite some evidence for the possibility that both human memory and the memory of the universe are stored in a depository beyond the physical universe, this assumption remains mostly hypothetical.

And yet, the hope for immortality lives on and keeps providing inspiration for scientific and non-scientific studies. Why is it the case? Because hope takes its strength not from knowledge, but from passion and belief. And most people passionately want to live for eternity and, subconsciously, believe in the supernatural. This subconscious belief feeds rational people's hopes for personal immortality.

A lonely traveller on a yacht in the middle of a boundless ocean – this is my image of a person at the dusk of his or her earthly journey who firmly believes in science and is unable to accept a common religious faith. The joys of the discoveries of childhood, the aspirations of youth, the confident steps and achievements of adulthood – all of these are behind. And what is ahead? Almost nothing, in all honesty. Of course, there still remains the health in the body, canned food in storage, fresh water in the cistern and the yacht is still afloat... But in the mist that lies ahead, the inevitable moment is already visible of when the journey will come to its end. And still, something elusive and subconscious glimmers in the darkness – that little ray of hope. The hope that the omniscient being is looking down at you and is ready to open their embrace to you, selflessly and unconditionally. The hope that your life was not just a splash on the surface of the boundless ocean of time, and your deeds – both good and bad – will be appreciated in eternity. For a non-religious person, this hope has no scientific or logical grounds. A person like that understands that by his or her conscious disbelief in God and the supernatural, he or she does not deserve this hope. And still, the hope is there. Locked in the dungeon of the subconscious, this hope feeds endeavours that, from a rational point of view, are impractical, and initiates enquiries into issues that are beyond our understanding. Against all odds, we live as if our soul is immortal.

But life itself is against all odds. There are no grounds for a system of such complexity as a human organism to exist and not to fall apart at any time, yet this system lasts. There are no grounds for a person to be kind to others, yet some people are kind. And it seems to me that this hidden, subconscious and 'unlawful' (for a rational person) belief that life and the mind are irreducible to the laws of nature, that there is something bigger than what a person and even humankind can possibly comprehend, that the supernatural is not a fallacy but a reality, is the feeding ground for the person's last hope: the hope for IS.

Notes

1 Winner, M. (2005)
2 http://en.wikipedia.org/wiki/Human
3 http://en.wikipedia.org/wiki/Future_of_the_Earth
4 https://en.wikipedia.org/wiki/Cosmic_Calendar
5 http://en.wikipedia.org/wiki/Human_extinction
6 http://en.wikipedia.org/wiki/Ultimate_fate_of_the_universe
7 Moravec, H. (1988)
8 Tipler, F. (1994)
9 de Chardin, P. T. (1961)
10 Krupp, E. C. (2003)
11 https://en.wikipedia.org/wiki/Thales#Beliefs_in_divinity
12 Subbotsky, E. & Trommsdorff, G. (1992)
13 Hutton, R. (1999)
14 Pronin, E., Wegner, D. M., McCarthy, K. & Rodriguez, S. (2006)
15 Subbotsky, E. (2001)
16 Subbotsky, E. (2005)
17 www.iep.utm.edu/immortal/
18 Dawkins, R. (1976)
19 www.goodreads.com/quotes/654503-that-man-is-the-product-of-causes-which-had-no
20 https://en.wikipedia.org/wiki/Astrology
21 https://en.wikipedia.org/wiki/Hellenistic_astrology
22 Gauquelin, M. (1969)
23 Gauquelin, M. (1991)
24 Eysenck, H. J. (1975)
25 Mayo, J., White, O. & Eysenck, H. J. (1978)
26 https://en.wikipedia.org/wiki/Mars_effect
27 Dunne, B. J., Nelson, R. D. & Jahn, R. G. (1988)
28 Jahn, R. G. (2001)
29 Jahn, R. G. & Dunne, B. J. (2008)
30 Subbotsky, E. (2013)
31 Subbotsky, E. & Ryan, A. (2014)
32 Moody, R. (1975)
33 http://en.wikipedia.org/wiki/Near-death_experience
34 http://en.wikipedia.org/wiki/Ancient_astronauts
35 Gray, J. (2007)
36 Kaku, M. (2008)
37 http://ru.wikipedia.org/wiki/Плутон
38 https://en.wikipedia.org/wiki/David_Hume
39 https://en.wikipedia.org/wiki/George_Berkeley
40 https://en.wikipedia.org/wiki/Arthur_Schopenhauer
41 Subbotsky, E. (2000a, 2000b)
42 https://en.wikipedia.org/wiki/Creation_science
43 https://en.wikipedia.org/wiki/Continental_drift
44 Carter, B. (1974)
45 http://en.wikipedia.org/wiki/Anthropic_principle
46 Barrow, J. D. & Tipler, F. J. (1986)
47 http://en.wikipedia.org/wiki/Copenhagen_interpretation
48 Wheeler, J. A. (1990)

49 http://en.wikipedia.org/wiki/Memory_consolidation#Synaptic_consolidation
50 https://en.wikipedia.org/wiki/Meno
51 Day, J. M. (1994)
52 Slater, A., Morison, V. & Rose, D. (1982)
53 Gibson, E. J. & Walker, A. S. (1984)
54 Baillargeon, R. (1987)
55 Onishi, K. H. & Baillargeon, R. (2005)
56 Sheldrake, R. (2013)
57 http://en.wikipedia.org/wiki/Joseph_Banks_Rhine
58 https://en.wikipedia.org/wiki/Verbal_overshadowing
59 Schooler, J. (2011)
60 Lehrer, J. (2010)

6
RELIGION AND BELIEF IN THE SUPERNATURAL

Problem

Moscow, January 11, 2014. It is bitterly cold. An immensely long line of people is entering the Cathedral of Christ the Saviour to venerate the Gift of the Magi brought from Greece for a few days. The loudspeakers in the underground stations ask people to postpone their visit to the Cathedral due to overcrowding, but in vain. The Gift is shown on TV: it is a small casket encrusted with precious stones and with an image inside. A never-ending flow of people approach the casket, kiss it or touch it with their foreheads, cheeks, hands or small icons. A reporter interviews some of the pilgrims. Many arrived from remote corners of Russia. All of them hope for a miracle: healing from diseases, happiness for themselves and family. According to legend, this relic has existed since the time of Christ. Like other relics, the Gift of the Magi is a gate linking the real world with the world of the supernatural – with god and angels. Perhaps 30,000 years ago, in the time of the Upper Palaeolithic, under the flickering light of burning pine branches, people touched the Spirit of the Great Deer pictured on the wall of a cave, asking the spirit to give them luck in hunting. After tens of millennia, the cave became the Cathedral clad in marble, and the image painted in ochre on the cave wall turned into a precious piece of jewellery, but human beings remain the same, with their pains and desires, love, hope and belief in the world that is invisible but can see them and help them if properly asked.

One day, while roaming YouTube, I came across a talk by Richard Dawkins – a militant atheist. Someone from the audience asked him why he was so sure that God didn't exist. "Because," he answered, "there is no evidence whatsoever of God's existence." And a memory crossed my mind of the early Christian author Tertullian (155–240 CE), who coined a phrase that could be summarised as 'I believe because it is absurd'[1]. Unlike Dawkins, Tertullian didn't think that the belief in God's existence should be based on evidence. Indeed, if there were evidence of God's existence, belief in God would not be necessary. Although we often say, "I believe in what I see with my own eyes," we don't really need to believe in something we have available in perceptual experience. For instance, knowledge about the law of gravitation is supported by our perception of weight, which provides us with evidence convincing enough to render the belief unnecessary. If someone doubted this law, the doubt would fast dissipate as soon as the person jumped from their chair and tried to fly in the air. Belief is different from knowledge. For belief, we need to imagine something that we can't possibly hold in perception and invest this something with existence by a sheer effort of will. But in order to make ourselves perform such an energy-consuming feat, we need to have a really solid reason.

But is this really true that there is no evidence of God's existence whatsoever? Can one live without any beliefs at all and rely only on scientific proof? Let's find out whether recent research on magical thinking and magical phenomena can help us answer these questions.

The discovery of the world of spirits

A vast herd of wildebeests slowly moves under the burning sun on the dry African plain. For many days, without water and food and crossing crocodile-infested rivers, the animals move out of the drought zone towards a zone with an abundance of green grass. Predators, which are territorial and stay in the drought zone, are hungry and thirsty as well, yet have to wait patiently for the return of the herds. But neither the herbivores nor the predators try to pray to gods for the desperately wanted rain. Humans are the only animal species to have invented prayer and ritual – ways of affecting nature by thoughts and symbolic actions. The belief that thoughts, words and rituals can directly affect natural processes and inanimate objects is the key feature of the belief in magic. This belief was also the first form of religion.

When did this belief begin? Some anthropologists associate the belief in magic with the neocortex, which evolved around 500,000 years ago and occupies a notable proportion of the human brain. The neocortex opened up the possibility for new cognitive skills to develop, which were necessary for the emergence of consciousness, language and religion[2]. Anthropologist Pascal Boyer suggested that the basis for the emergence of religion was laid down by the evolving of such cognitive skills as attribution of activity to natural processes, attribution of consciousness to animals and inanimate objects and some others. He writes, "When we see branches moving in a tree or when we hear an unexpected sound behind us, we immediately infer that some agent is the cause of this salient event. We can do that without any specific description of what the agent actually is"[3], p. 144.

One can hardly argue against the idea that a belief in invisible agents that are hidden in rivers, boulders, trees and animals requires a complex brain and advanced cognitive skills. The question, however, is whether having a complex brain and advanced cognitive skills was sufficient for humans to develop the idea of invisible agents. The precursors of such cognitive skills exist in apes and other animal species, but these animals did not invent religion. There is a fundamental difference between *danger-escaping behaviours based on detecting movements and sounds in the environment* and the animistic belief that *behind these movements and sounds some invisible agent is pulling the strings*. Danger-escaping behaviours can be either hardwired into animal brains by evolution (and then they are called instincts) or learned through experience (conditioned reflexes), but they are not based on beliefs. In order to believe in invisible agents, spirits and gods, people had to first develop the idea of what a spirit is. In other words, cognitive skills, however complex and advanced, are nothing but intellectual tools, just like a stone axe or a knife. But the idea of an agent or a spirit is an abstract notion that required a leap from the world of perceptual experience into the invisible world of abstract thinking. A leap like that needs two things: a powerful imagination and a strong emotional experience of some kind.

As argued in Chapter 2, the emotional experience that triggered the invention of the concept of a spirit was the realisation of personal death. Dreams and hallucinations could be additional factors that helped to develop the notion of a spirit. In their dreams at night, people had experiences that differed from those they had in a waking state: they could fly in the air, go through hard walls and speak with their deceased relatives. These experiences may also have contributed to the idea that in a person's body an entity lives that can go out of the body and wonder around while the person is asleep. Taking this possibility on board, British anthropologist Edward Tylor maintained that the ancients mistook their dreams for reality. This 'error of judgement' created a belief in the invisible world of spirits[4]. What is missing in Tylor's account is that in order to make this error of judgement, people had to first notice that they have dreams. Neurological studies suggest that some animal species have dreams[5], but animals are unlikely to ponder their dreams. Pondering one's dreams requires reflective thinking – looking at dreams that are not currently present but need to be recalled from the perspective of a waking state of mind. Making the leap from the perceptual world of the present into the invisible past required a push from some powerful emotional experience. The shock at the realisation of the inevitability of personal death may have been such a push. Having invented the idea of the spirits of the dead, people could assume that living persons also have spirits inside their bodies and that these spirits travel outside their bodies in their dreams.

In sum, the discovery of the magical world of spirits was triggered not by intuitive cognitive skills and logical inferences, but by the powerful desire to live and the fear of death. In modern people, the longing to stay alive and the fear of death are still there. But today, due to the achievements of science and medicine, people live longer and are less dependent on the unpredictable circumstances of life. Many educated rational adults reason that magic and God are concepts from the genre of fantasy and don't really exist. But do they really believe this?

Belief in the supernatural

For a child brought up in a religious family, a belief in the supernatural comes from his or her social environment. The child absorbs their parents' religious views and rarely doubts them. In a similar way, adult religious believers seek support of their faith in their co-religionists. Psychological experiments have shown that, for most people, even a belief in science, just like a belief in God or in magic, is based not on independent experimentation and critical thinking, but on the support of their trusted social groups. For example, in one experiment, educated adult participants (university undergraduates and staff members) were individually shown an empty wooden box, asked to place a brand new plastic card in it and close the lid of the box[6]. Next, the experimenter chanted a magic spell intended to damage the object in the box. The participants were then asked whether they believed the object in the box had changed; most of them answered that they didn't. The participants were instructed to open the box and remove the card. They were surprised to see that the card now had engravings on it as if done by a sharp instrument. The participants were encouraged to examine the box for a false bottom and other hidden compartments, and found nothing. Despite the absence of a rational explanation, most of the participants denied that the card had been damaged by the magic spell. They argued that changing physical objects by just saying words is in contradiction with the laws of physics. Next, the experimenter asked the participants to put their hands in the box and said that he did not guarantee the safety of their hands if he said his magic spell again. This manipulation aimed to remove the implicit assumption that the laws of physics are immutable. The experimenter also asked the participants to give written consent that they would not blame the experimenter for whatever happened to their hands after he said his magic spell. This was done in order to remove the participants' feeling of safety for their hands, which would usually be guarded by the conventional moral rule of not hurting a participant during an experiment. The participants now understood that their belief that a magic spell could not possibly hurt their hands was no longer ensured by social conventions and hung on their personal courage. When, under these circumstances, the experimenter asked the participants' permission to repeat his magic spell, half of the participants denied this permission and explained that their decision stemmed from their belief that the spell might indeed damage their hands. This showed that *removing the implicit social support for the belief that the laws of physics are immutable made educated adult participants abandon this belief and admit that they actually believed in magic,* both in their verbal judgements and in their behavioural reactions. In subsequent experiments when participants had to put at risk not their hands but their future lives, up to 90% of participants acknowledged that magic could indeed work[7].

It is noteworthy that educated British participants showed a greater disbelief in magic than uneducated peasants in rural villages of central Mexico only when the implicit social support for this disbelief was provided. However, when this social support was removed, British participants exhibited a belief in magic to approximately the same extent as Mexican participants[8]. These data explain why, in the ancient world, in Medieval Europe and even in the time of the Renaissance when a belief in magic was supported by the dominant social ideology, this belief was overwhelming. It is only because of the phenomenal success of science in the last four centuries that the dominant social ideology changed and the belief in science took over. Yet the aforementioned experiments reveal that modern rational adults who consciously deny their belief in magic subconsciously still harbour this belief. The question arises as to why the belief in the supernatural is able to withstand the pressure of scientific education. And why is religion still here? Could it not be because there is some evidence for the existence of God and the supernatural?

Evidence for the supernatural

Paradoxically, the idea of the existence of a supernatural, omnipotent intelligence comes from science itself. We know that mathematics and logic operate on 'borderline concepts', such as 'infinity'. But how could people possibly acquire such concepts and use these concepts in scientific research? Certainly, they could not learn the infinity concept from experience, since in experience we only see finite things. For example, when we think of the concept 'a bird', we can show a real bird flying in the sky in support of this concept. When we think of a mathematical concept, such as 'number 7' or a triangle, we can point to a set of seven apples or the triangular-shaped facet of a crystal. By contrast, when it comes to borderline concepts, such as infinity, there is nothing out there to show in support of it. It is impossible to see real-life infinity; the concept of infinity can only be represented as a symbol ∞. But we know that infinity is not just a symbol, it really does exist, because if it didn't, we wouldn't know about it. Of course, there may be other characters that are only imaginary and don't have referents in the real world, such as mythological creatures (e.g., the Minotaur, a creature half like a man and half like a bull), but these characters are still based on our experience – seeing a man and a bull and simply putting parts of these creatures together in the imagination. Not so with infinity. Infinity couldn't originate in the human mind, since the human mind is only capable of performing a finite (and not very large) number of operations. Even the most powerful quantum computers would be able to complete an 'almost infinite', yet still limited number of operations. Infinity cannot exist in the physical universe itself, as infinity is an infinite sum of mental operations, like the sum of an infinite set of real numbers, and the physical universe is devoid of any mentality. We have to admit on this ground that infinity exists within some kind of omnipotent mind – the mind of God – that simply shared its knowledge with us.

Usually, scientists are not bothered by this sort of philosophical problem; they simply use the infinity concept to find solutions to many theoretical and practical problems. But the effectiveness of borderline concepts is the indirect proof that the 'holder' of these concepts – God – exists as well. Of course, this 'god of science and philosophy' isn't exactly like the gods of the ancients, nor even like the God of modern religions, yet there are some similarities. The god of science has a mind that is infinitely more powerful that the human mind. Some philosophers argue that we will never be able to understand the mind of God[9,10]. But one doesn't have to understand the mind of God in order to infer that God exists.

Another kind of indirect evidence for the existence of the supernatural comes not from theoretical considerations, but from phenomena that cannot be explained by science. The Big Bang that brought the universe into being, the presence of the enigmatic creative force in the universe that builds complex systems out of simple ones, dark energy and dark matter, the emergence of life from inanimate matter, the emergence of consciousness and free will – these and other phenomena have thus far escaped scientific understanding. Of course, science will keep trying to reach a better understanding of such phenomena, but it is unlikely that science will ever be able to fully explain them. And this means that there is a mysterious 'point of creation' within or outside the universe that is beyond reach of the human capacity for rational comprehension.

But let's come back to the point in early prehistory when the belief in the supernatural emerged. People discovered the fact of their mortality. This discovery brought them to the idea of an afterlife and the notion of a spirit or an agent. The invisible world of spirits expanded to include living humans, animals, plants and inanimate objects. The whole world began to be viewed as being full of agents and spirits. The power of the gods grew from looking after living tribesmen to commanding the elements – air, water and fire. Still later, in monotheistic religions, a single god became the maker of the world; from being a god of only one people, he became a god for all people on earth. Eventually, people's view of the world changed as well: the scope of entities with spirits inside shrunk to include only God, people and, arguably, animals. There appeared a vast area of spiritless objects governed by unchangeable and universal laws. Physical science emerged, and then other natural sciences that model themselves on physical science. Science, technology and medicine undermined the leading role of religion in modern industrial cultures and ousted the belief in magic to the realm of the subconscious. But science failed to completely replace people's belief in the supernatural.

The question arises as to why the belief in the supernatural didn't vanish entirely in the course of historical development like many other beliefs disappeared, such as beliefs in mythical titans and that the earth is flat. The aforementioned evidence for the existence of the supernatural is indirect and unlikely to prevent most people from abandoning their belief in magic and in God. So, why did beliefs in the supernatural survive? A possible answer is: *because these beliefs serve people's needs that cannot be served by science.* In particular, the possibility of addressing gods through prayers and rituals helps people reduce their fear of death and suffering, thus increasing people's chances for survival. In psychology, this psychotherapeutic function of belief in the supernatural is called 'the illusion of control'.

The illusion of control

Psychological experiments have shown that people feel more confident and act more effectively if they believe that they are in control of a situation than if they think that the situation doesn't depend on their actions. For example, in one experiment, a group of participants was allowed to choose their lottery tickets; another group of participants received lottery tickets chosen for them by someone else. Although both groups had equal chances of winning the lottery, participants from the first group assessed their chances as higher and were inclined to swap their tickets for different ones to a lesser extent than participants of the second group[11]. In another experiment, participants were instructed to watch a basketball player throwing a ball into a basket. Those participants who had been trying to visualise successful throws and thus 'help' the player to score did indeed think that the player's successes were achieved partly due to their imaginary 'help', despite the participants' realisation that their imagination could not possibly affect the player's actions[12]. One can assume on this ground that a petitionary prayer and other ritualistic pleas to gods play a similar role: they create in people an illusion of control over those situations of their lives that, in reality, are out of their control.

Indeed, why are many people so eager to look into their future by addressing oracles and other fortune-tellers? Logical reasoning should tell people that if something were written into their fates, then it would happen to them independently of whether they know or don't know their fates. Yet people keep trying to learn their fates. In ancient Rome, no serious actions (like going to a war or laying the foundations of a major building) were taken without consulting fortune-tellers and studying omens. Even the decision to convert to Christianity was taken by the emperor Constantine because of the omen he had seen before a battle. Today, people are less superstitious, yet fortune-telling is still a popular profession. In the media and on the web, a small army of fortune-tellers predicts actions of terror, earthquakes, tsunamis and other disasters. What can science do to oppose this mania of prophesising? This is not a realm for science. Even religion prefers not to intervene. As a result, the role of Nostradamus is up for grabs, and there is no shortage of candidates. So, why do rational people seek to know their fates?

The 'illusion of control' phenomenon might suggest an answer. People want to know their fates in order to feel they are in control of them. They understand of course that their attempts to change fate are futile and illusory, and yet this illusion has a soothing psychotherapeutic effect. In his poem 'The Song of Wise Oleg', Russian poet Alexander Pushkin shows us an example. Prince Oleg asked a fortune-teller when and why he would die. The fortune-teller prophesised that Oleg would take death from his favourite horse. Oleg was sceptical about the prophecy, yet he replaced his stallion with another one. Many years passed. Oleg successfully fought and won many battles. One day, Oleg enquired about the fate of his favourite horse and was told that the stallion had long been dead. Oleg mocked the false prophecy and decided to visit the horse's grave. When he put his foot on the horse's skull, a poisonous snake slithered out of the skull and bit Oleg. Oleg died, and the prophecy came true. Oleg's attempt to cheat fate failed – and yet it was not entirely futile. Let's imagine that Oleg had not changed his horse. In every battle, he would have been worried that the stallion was destined to bring him death, and this anxiety may have distracted Oleg, diminished his courage in the battle and even cost him his life. Changing the horse increased Oleg's confidence and supported his military spirit. It didn't matter that in the end the prophecy came true. What mattered is that the illusion of having cheated fate gave Oleg a very real advantage: courage and self-confidence in battles.

For believers in God, religion provides a feeling of control. Through prayer and confession, our sins can be obliterated; for a religious person, even death can be taken under control, for there is hope of resurrection. For those who don't believe in God, everyday superstitions perform a similar function of creating an illusion of control. By crossing our fingers or knocking on wood, we make ourselves calm, as if pacifying evil forces and soliciting the help of our guardian angels. We know that these actions are nothing but superstitions, and yet these magical actions help. It is no coincidence that people of most dangerous professions (pilots, construction workers, sportsmen) are also the most superstitious[13]. Where results are unpredictable and the stakes are great, science cannot help. All that remains is to rely on the supernatural, even if a person does not believe in it – more so if he or she does.

"But why do so many people today regress to the level of ancient magic instead of going with religion?" a reader may ask.

Magic and religion

As argued in previous chapters, initially magic and religion were the same. People believed in ancestral spirits and animal spirits and those were their gods. Eventually, in great civilisations like ancient Egypt, Greece and Rome, the gods acquired more diverse and clear forms. But along with the main gods, there always existed a multitude of little gods and goddesses of the second rank, with which people communicated by means of magical rites. With the onset of monotheism, tension developed between the dominant religions (e.g., Judaism or Christianity) and magic, for they started compete for people's minds.

Indeed, the Bible condemns magic. "Never let a witch live" (Exodus, 22:18) is just one of the many places in the Old Testament where magic is ostracised. And yet both the Old and the New Testaments are full of miracles. In the Old Testament, God creates the universe, including plants, animals and humans (Genesis, Chapters 1–2); God sends the Great Deluge (Genesis, Chapters 7–8); at Tower of Babel, men's single language changes into many (Genesis, 11:1–9); the sinful cities Sodom and Gomorra are destroyed by heavenly fire (Genesis, 19:24); the wife of Lot becomes a pillar of salt (Genesis, 19:26); Aaron's rod is turned into a serpent (Exodus, 7:12) – these miracles constitute just a small part among many other such instances. In John's Gospel, at a wedding in Cana, water is turned into wine (2:1–11); a royal official's son is miraculously healed (4:46–54); during the autumn festival in Jerusalem a man is healed of a 38-year infirmity (5:1–9); Jesus heals a man born blind (9:1–7); and Lazarus is resurrected after being dead for three days (11:38–44). In Matthew's Gospel, Jesus casts demons out of people with a word; and finally, Jesus himself is resurrected from the dead.

So, the Bible doesn't deny that magic and miracles do really exist. Even today, the Catholic Church believes that during Eucharistic Celebration, the sacrificial bread and wine become the body and blood of Christ through consecration by a priest. In this rite, bread and wine do not simply symbolise the flesh and blood of Christ, but undergo *transubstantiation* – the magical conversion of the invisible substance of the bread and wine into the flesh and blood, while the sensual appearances of the bread and wine remain unchanged[14]. From a historical perspective, this isn't surprising: it is impossible to cut the umbilical cord that links religion to its mother – magic – without spilling some blood in the form of miracles. From the psychological perspective, the Bible needs miracles to be convincing to the eyes of believers. Without miracles, the Bible would not differ from an ordinary, and rather messy, description of historical events and historical anecdotes, and God Almighty and his representatives on earth would not be able to inspire awe and enjoy the unconditional trust of people. It appears that in the view of the Bible's authors, doing miracles as such is not a sin; a sin is when a person tries to do miracles without the help of God. Considering that believing in magic is believing in spirits that live in the things of nature, doing magic without God's authorisation means making a pact with the little spirits while failing to ask the permission of the supreme spirit – God – at the same time. Because priests represent God on earth, doing magic also undermines the priests' authority. It is unacceptable for the Church to share magical powers with commoners; doing so could diminish the priests' psychological influence over the people.

As a result, magic split into two categories: high magic and low magic. Initially, the belief in magical reality was a way of overcoming the fear of death and anchoring a person's existence, which is full of suffering and unpredictable dangers, to the everlasting world of the spirits. Magical communication with gods pursued fairly practical goals: people asked spirits to give them luck at hunting, crop growing or war. With time, in monotheistic cultures, this role was taken by official religions, and the remaining part of magic turned into illegal 'direct deals' with little spirits. Many of those who go to practicing witches believe in God, but they see God as a powerful wizard of some sort who is high above and hard to reach. By contrast, the 'little spirits' look like more accessible partners in business; they are less strict and can be bribed. Even within the realm of the main religions, in people's daily lives, the 'high' and 'low' requests for miracles are erratically intertwined. Official religious ceremonies on Christmas and Easter are one thing, but in daily life people ask God for simple and mundane favours. Recently, I visited a monastery near Moscow. The monastery's role as a place of worship is accompanied by a set of practical activities, such as selling soft drinks, cakes, calendars and icons. In the cathedral, people habitually cross themselves in front of the icons, light candles and queue for the holy water; the hands and legs of a monument to the saint who is buried in the monastery's cemetery are polished to shining by the touch of thousands of believers who have sought the benevolence of divine forces. Magical rituals on this ground are as mundane as consuming tasty cakes stuffed with cabbage.

And yet, in industrial cultures, the number of people who do not belong to any established religion grows. A number of factors may have contributed to the growing secularisation of industrial societies. As argued above, one of the factors was the success of science, which made human life safer and longer. Another factor is the stiff nature of religion itself. Mainstream religions like Christianity are based on certain fundamental assumptions (e.g., the immaculate conception of the Virgin Mary, the resurrection of Christ after death), which for many modern educated individuals are hard to comprehend and accept. At the same time, the modern Church is reluctant to accept real supernatural phenomena, such as parapsychological effects, on the grounds that such phenomena may be inspired by dark forces. Finally, the abundance of grief and unhappiness in the world, the observation that evil is not always punished but often is rewarded, stand in contradiction to the image of God as a kind and benevolent entity. All of this diverts many people from the established Church.

Does this mean that these people live without beliefs at all? Or is it more likely that when the energy of a belief in the supernatural is left without its main target – God – it finds a replacement? The new target could even be the theory that a belief in God is harmful and dangerous for people. There are atheists who fight religion with the zeal of a religious extremist, as if God himself confirmed their disbelief in God. More often, however, the disappointment in mainstream religions makes people search for alternative ways of dealing with the supernatural. One of these ways is applying to traditional magic and superstitions, and another is *rational mysticism*.

The belief 'from reason'

In one of his books, American psychologist William James (1842–1910) described the unusual sensations he experienced after taking chloral hydrate and other hallucinogenic drugs[15]. Although James was a firm believer in the scientific method in psychology, he nevertheless admitted that mystical experience could be a reality. In his view, mystical experience allowed a person in certain moments of life to switch to the 'upper reality' – to the 'mind of God'. Like James, some modern researchers believe that one way to achieve such an altered state of consciousness is by taking psychedelic substances.

In 1954, English writer Aldous Huxley reported his experiences after taking mescaline – a psychedelic drug obtained from peyote cactus[16]. Huxley revealed that he had an experience of oneness with a higher reality, which subsequently positively affected his life by broadening his consciousness and increasing his interest in life. In order to verify Huxley's report, in 1962, psychiatrist and theologian Walter Pahnke conducted an experiment in Harvard University that became known as the 'Good Friday Experiment'[17]. Participants (university undergraduates) were divided into two groups. In the experimental group, participants received a pill of psilocybin (the active ingredient in psilocybin mushrooms) that produces altered states of consciousness, and the control group received an 'active placebo' – a dose of niacin that produces some physiological changes but doesn't affect consciousness. The participants were asked to describe their experiences and assess the strength of those experiences twice: immediately after the session, which lasted for a few hours, and six months later. The participants who received psilocybin, but not the control group, reported profound mystical experiences, such as a feeling of oneness with surrounding objects, of the divinity and uniqueness of each object, of the impossibility of putting their experiences into words and of the loss of awareness of their location in space and time. Six months later, the participants acknowledged that the mystical experience had had a positive effect on their lives: it had made their religious beliefs more profound, intensified their feelings of compassion and love for others and sharpened their perceptions on the beauty of life. This experiment was taken by many as proof that an artificial intrusion in brain chemistry could produce mystical experiences similar to those achieved by traditional meditation practices, such as prayer, concentration or long fasting. However, not everyone agreed.

Harvard university graduate Rick Doblin interviewed the participants of this experiment in 25 years after the experiment and found that, along with the positive experiences, they'd also had negative ones, such as feelings of fear, anxiety and concern that they were going mad. One of the participants who had taken psilocybin fell into the state of paranoia, rushed out into the street and started to proselytise the coming of the age of global peace. Even after 25 years, this participant occasionally felt fits of unaccountable anxiety. In other words, mystical experiences induced by psychedelic substances can be both positive and negative. Philosopher Steven Katz considers mystical experiences achieved by taking drugs as false. In his view, such drug-induced experiences are nothing but a reflection of noise in the brain of the person who took the drug, whereas meditation, yoga and prayer could indeed bring the person into contact with a higher reality[18].

With the aim of studying this issue in depth, American radiologist Andrew Newberg and psychiatrist Eugene d'Aquili used the method of photon emission tomography to explore the brain activity of people who were at the peak of meditative states[19]. They discovered that when people are immersed in a deep meditative state and experience a feeling of oneness with the universe, their prefrontal cortex becomes hyperactivated, which leads to the inhibition of neural activity in parts of the parietal lobes. Newberg and d'Aquili hypothesised that it is the inhibition of the brain activity in the parietal lobes that causes some of the characteristic phenomenological features of mystical experiences. They assumed that with the development of new advanced technologies for registering brain activity, it would be possible to experimentally determine whether a mystical experience is genuine or results from pathological conditions, like hallucinations in schizophrenic patients. The authors cited studies that had shown that having a genuine mystical experience positively affected people's lives by increasing their self-esteem, improving interpersonal relationships, decreasing anxiety and enhancing the ability to feel compassion for other people[20]. However, further research revealed that mystical experiences initiated by psychedelic substances could also lead to increases in narcissism, fanaticism and hatred.

If Newberg and d'Aquili thought that they had found the brain localisation of one particular kind of mystical experience – the feeling of oneness with the universe – then Canadian psychologist Michael Persinger claimed that he had found the localisation of all kinds of mystical experiences, including the belief in God[21]. He reported that many of his patients with brain trauma had experienced a feeling of some kind of ethereal presence, as if someone invisible was near them. If the right brain hemisphere was damaged, patients felt that a demon or a ghost was chasing them. If the damage was in the left hemisphere, then the feeling of a presence took the form of voices. People with damage to the left temporal cortex were particularly prone to this kind of experience.

Persinger's findings had precursors. It had been established that epileptic fits are linked to disturbances of brain activity in the left temporal cortex where the functions of speech and 'feelings of personal identity' are localised. During epileptic fits, people may experience the feeling of the 'presence of God'. Some theorists suggest that prominent historical characters, such as the prophet Ezekiel, Joan of Arc and others, suffered these kind of fits. Relying on this and other evidence, Persinger hypothesised that the left temporal cortex, which gives a person a feeling of 'oneself', when disturbed begins to interpret signals that come from the person's right hemisphere as the presence of 'another self'. Depending on a person's life history, the person may interpret this 'other self' as an alien, a ghost, an angel or a god. To examine this hypothesis, Persinger and his colleague Stanley Koren created a special device they called the 'Octopus', which later became known as the 'God helmet'. The device is a helmet consisting of eight electromagnets capable of emitting weak electrical fields; when placed on a person's head, the device can stimulate various parts of the cortex[22]. The scientists reported that under certain spatial–temporal patterns of stimulation of the temporal lobes, up to 80% of participants had the 'presence' experience. However, Persinger's interpretation of the feeling of 'presence' as the presence of supernatural beings is questionable. It is possible that what was present was an alter ego of the patient himself or herself. Earlier experiments by Canadian neurosurgeon Wilder Penfield showed that electrical stimulation of the temporal cortex could induce subjective experiences in epileptic patients that felt authentic[23]. This suggests that the 'presence' phenomenon evoked by the 'God helmet' could be an artificially induced simulation rather than a genuine mystical experience.

Interestingly, the 'presence effect' enhances the ability of extrasensory perception (ESP); 75% of participants who wore the God helmet guessed telepathically transferred messages correctly, whereas the probability of correctly guessing by chance was only 20% (see the next section). The assumption that there is a connection between the 'presence' experience and ESP is not arbitrary. Among many of God's supernatural abilities is the ability to read people's minds. If stimulation of a certain area in the brain connects a person to a higher reality, then this reality might indeed enhance a person's supernatural abilities, such as ESP and telekinesis. If it were possible to show in experiments that these extraordinary abilities are real, this would be more evidence for the existence of the supernatural.

Psi as a porthole into the supernatural

Why do stage magicians perform tricks? Why do we enjoy watching the tricks while we know that these are just illusions? Could it be because at the bottom of our minds we still harbour the belief that the impossible may be possible? We know that a human being can't levitate in the air or walk on water, but we can see ourselves doing these things in our dreams. In our dreams and the imagination, all sorts of impossible events occur, although our rational mind marks the existence of these events as only ephemeral. And now suppose that we see something impossible, like a person walking on water, happening on the stage. Immediately the thought "a person can walk on water" is upgraded in its rank on the scale of existence. Just for a fleeting moment at the back of our minds, the miraculous phenomenon that we are seeing becomes a reality, even while our critical thinking keeps saying it is a trick. Seeing the miracle on the stage evokes in us the emotion of surprise and, usually, happiness[24].

The emotion of surprise is understandable, but why happiness? Perhaps because seeing a supernatural phenomenon touches upon the hope hidden deep in our subconscious that the supernatural is indeed real and we can be a part of it. If miracles do exist, then a miracle could happen to me, and my life would not be just a random splash on the boundless ocean of time, but instead has a meaning! Experiments have shown that both preschool children and adults, equal in all other conditions, prefer watching a magical event to watching a novel physical event[25]. As children, we like playing magical games of pretend, because the feeling of being a powerful wizard or fairy helps us overcome a feeling of inferiority and gives us a feeling of control over the external world (see Chapter 8). In their attraction to magical thinking, adults are no different from children.

In essence, stage magic is a simulation of the supernatural abilities of the mind. In psychology, stage magic can be used for studying cognitive abilities, such as perception, thinking and memory[26]. In my own research, I have used magic tricks to study magical thinking and beliefs in supernatural phenomena[27].

But could supernatural phenomena themselves be studied under strictly controlled conditions, like experiments in science? Some scientists decided that they could. Thus emerged *parapsychology – enquiry into the supernatural phenomena by scientific means.*

Parapsychological phenomena have been under investigation for over 100 years. One such phenomenon is ESP – the ability to read other people's thoughts at a distance or foresee the future. A typical such experiment uses Zener cards – five cards with one iconic geometrical shape (like a cross, a circle or a star) printed on each card, with five copies each. In a test for ESP, a person (an inductor) picks a card out of the shuffled pack and, looking at the picture, tries to pass the image to another person (a receiver) who is situated in another room or building. The timing of the information transfer and reception is synchronised in advance. The receiver tries to guess and draw the image. If the number of correctly guessed shapes is higher than that expected by chance alone, then the result is viewed as confirming that the information transfer did indeed take place. There are multiple versions of this method. Many years of research into ESP have shown that the effect is small but statistically significant[28]. Another effect studied by parapsychologists is telekinesis – the ability of a person to affect certain physical processes by a sheer effort of will. This effect, too, was supported in rigorous experimental studies[29,30]. Nevertheless, most mainstream psychologists are not persuaded by these results.

As a researcher who had never ventured in parapsychology, I too had my doubts about the reality of parapsychological phenomena, and so I decided to conduct a few experiments myself. Because such experiments are time and labour intensive and the chances of their publication in a mainstream journal were near zero, my attempts were motivated by sheer curiosity, and I conducted the experiments in my free time. My first attempt was made in cooperation with a researcher of anomalous phenomena, Adrian Ryan[31]. The experiment was on ESP, and we used the method developed by American physicist and parapsychologist Edwin May. A participant is told that in approximately 15 minutes he or she will be shown a photographic picture of a certain landscape or a city. What exactly this picture (named the 'target') will show nobody knows, because the picture is on a computer's hard disc mixed with 300 other pictures. The participant's goal is to visualise and draw this picture. The participant is instructed, after the experimenter gives a signal by saying 'target', to make a sketch with a pencil on a sheet of paper of the first image or landscape that crosses the participant's mind. After the sketch is made, the experimenter switches the computer on and the computer randomly selects five pictures out of the pool. It is assumed that the target is among these chosen pictures. Next, looking at the participant's sketch, the experimenter ranks the five chosen pictures on a scale of similarity to the participant's picture, with the least similar being given the lowest rank and the most similar the highest rank. The computer is then put to work again and randomly picks a picture out of the five previously chosen pictures. The chosen picture is considered to be the target that the participant was trying to see by making his or her sketch. The experiment ends with the experimenter showing the target picture to the participant.

The experiment's point is as follows: if the target picture chosen by the computer out of the five pictures is the one that the experimenter had ranked the highest, then the experimental trial is viewed as a 'hit', meaning that the participant correctly guessed the picture from the future; if, however, the target picture is ranked low in similarity with the participant's drawing, then the trial is considered a fail. If the participant was not able to 'see the future', then the experimenter's highest-ranked picture would be chosen by the computer by chance alone; that is, in 20% of all trials. If, however, the participant could indeed see the image from the future, then the target picture would coincide with the experimenter's highest-ranked picture significantly more frequently than in 20% of all trials. It is clear that in this experiment we are dealing not with the ordinary anticipation of future events on the basis of our past experience. For example, when we go to a cinema to see a new blockbuster movie, we can anticipate the content because our past experience tells us how blockbuster movies are made and what they are usually about. By contrast, in this experiment, participants' past experiences are of no help, because past experiences cannot tell the participants what picture out of millions of possible landscapes will be shown to them.

Participants were divided in two groups. The participants from one group were rewarded for participation with a small sum of money, independently of their success at guessing the target picture. In the other group, participants were promised a sum 20 times larger than the normal reward if their trial was a hit. The idea was to find out whether an increase in motivation would help the participants from the second group mobilise their ESP ability and produce a larger number of hits than participants from the first group. The results were unexpected. The participants from the first group produced a number of hits that was significantly above chance, whereas participants from the second group scored at chance level. This indicated that participants from the first group, being rewarded with a small sum independent of success, indeed exhibited the ability to 'see the future'. Albeit their sketches were not always a mirror image of the target picture; sometimes the similarity was in meaning rather than in resemblance. For example, one participant drew a road, a windmill on a hill and mountains on the horizon, and the target picture proved to be almost exactly the same. Another participant drew a strange human figure wearing a helmet and said it was an alien. None of the five pictures chosen by the computer even remotely resembled the figure of the alien. However, I noticed than one picture showed a wheat field with concentric circles and ellipses, as if having been printed on the field by a gigantic stamp. I remembered that crop circles are sometimes viewed as being made by aliens and so gave this picture the highest rank; this picture indeed turned out to be the target picture.

There are various possibilities for interpreting the significant paranormal effects of this experiment, but all of them include a direct effect of Self on random events. One possibility is to see it as a result of the *experimenter's expectancy effect*. Since the experimenter knows the picture that the participant has produced, he could subconsciously affect the computer's random number generator and so influence the target selection. Such interactions between the Self and random target selection have been reported previously[32].

The question is why the participants who had been promised a large reward for successful guessing the target picture did not exhibit ESP. A possible explanation lies in the 'optimum of motivation' effect. According to this effect, if the achievement motivation is too great, animals and people start to become anxious, and the effectiveness of their actions goes down[33]. In a replication of the experiment, the used method was the same, but the reward promised to the second group for successful guesses was a lot smaller. The aim was to find out whether decreasing the motivation to succeed at guessing the target picture would help participants of the second group do better. However, this time neither of the two groups exhibited ESP.

A second study was based on the method proposed by American social psychologist and parapsychologist Daryl Bem. Bem claimed that his study showed the effect of 'retroactive facilitation of recall'[34]. The gist of the method is as follows: participants were individually presented with a succession of 48 words on a computer screen that named one of four types of objects: food, clothing, professions and animals. The participants had been instructed to visualise every word that they saw. After all the words had been presented, participants were unexpectedly given a memory test that asked them to recall as many of the words as they could. So far, this method represents a standard psychological test of memory. But then the most interesting part began. The computer randomly selected 24 words out of the previously presented 48 words, and the participants were instructed to classify these words into the four aforementioned categories (i.e., food, clothing, professions and animals). In other words, the participants were given a practice exercise – the opportunity to practice with these words and so better remember them. If this practice exercise had been given *prior to the memory test*, we would not be surprised to find out that the participants recalled the words from the subgroup that they saw twice (let us call this subgroup S2) better than from the subgroup of the remaining words, which they only saw once (S1). However, because in this experiment the practice exercise was given *after the memory test*, the participants were expected to remember an approximately equal number of words from both of the subgroups.

But the unusual result reported by Bem was that when the computer counted the correctly recalled words, it found that the participants recalled a significantly larger number of words from S2 than from S1. It appeared that the principle 'repetition is a mother of learning' worked backwards: even though the participants repeatedly saw the words from S2 *after they had recalled them*, these words were still remembered better than the words from S1. Understandably, in the *control session* of this experiment, in which participants were not given the practice exercise and the computer still randomly selected 24 words out of 48, there was no difference in recall between the two subgroups of words. Having published his study in a mainstream journal, Bem put a detailed description of his method on the Internet and welcomed replication. I was one of dozens of scientists who conducted replications. Most replications failed, but there were around a dozen successful attempts[35]. My replication differed from the other replications in that I repeated the experiment three times with two groups of participants (experimental and control) in each trial. I also conducted all of the experiments in person, whereas most other replicators passed the duty of data collection down to students or research assistants. In three out of six groups of participants, I found significant differences between the numbers of words remembered from S2 and S1. The probability of this happening by chance alone was 0.002, which was significantly lower than the criterion of 0.05 for the 'no effect' assumption that is accepted in science. This means that my results could not be explained by mere chance.

The difference between Bem's results and the results of my replication was that in two out of three trials that produced significant differences in recollection, the words from S1 were recalled significantly better than the words from S2. Also, in one of these trials, the significant effect was found in the control group and not in the experimental one. This goes against Bem's hypothesis that memory facilitation through practice works backwards, 'from the future to the past'. It looks more likely that my results in this experiment, like in the previously described experiment, were caused by the experimenter's expectancy effect – the *direct effect of the experimenter's expectation on the data of the experiment*. This conclusion, however, warrants some explanation.

The standard 'observer effect' is well known. This effect means that an experimenter can unintentionally influence the results of an experiment by giving the participants suggestive signals through voice intonation, facial expressions or by skewing the experimental data in the direction of the experimenter's expectation. Fortunately, my experiment was safeguarded against the standard observer effect because the participants could neither see nor hear the experimenter; during the experimental session, I was isolated from the participants by an opaque screen, and the participants received all of the instructions directly from the computer. The results were registered and statistically worked out automatically by the computer; the calculating program was locked, which excluded the possibility of tampering with the data. The only way for the experimenter to affect the results was if the experimenter's mind was employing its ability of telekinesis by directly influencing the random number generator in the computer. Because I already knew the results of Bem's original experiment, my expectations regarding the results of my experiment were biased by this knowledge; subconsciously, I expected that the words from S2 and S1 would be recalled with different success rates. Independently of my conscious intentions, my subconsciousness might have switched on and off the program of random numbers that the computer contained, so instead of randomly choosing the words out of the pool, the computer loaded the words that the participant had recollected into S2 more frequently than into S1, or vice versa. Another possibility was ESP – I might subconsciously access the information of which words would go into S1 and S2 and transfer this knowledge to the participants' minds, thus helping the participants remember words from one of these groups better than from the other. In a published paper[36], I called the results of my experiments the 'non-standard observer effect', which essentially is a case of a direct Self over matter/mind magical phenomenon (see Table 4.1).

My results contributed to the discussions in regard to the effects reported by parapsychologists. They showed that other researchers could replicate these effects, but not every time and not always in exactly the same form. This 'diminished replicability' distinguishes parapsychological effects from scientific ones, where full replicability is required. What could be a cause of this diminished replicability of parapsychological effects compared with the full replicability of effects in the physical sciences? In my view, the cause is hidden *in the nature of the objects studied by the physical sciences and by parapsychology*. Sciences like physics or chemistry study inanimate objects that are devoid of any spontaneity and unpredictability within themselves. For instance, the physics of solid states studies metals, magnets, superconductors and similar macro objects whose properties are always the same. By contrast, parapsychology studies the abilities of the human mind, such as ESP, telekinesis and clairvoyance. The human brain is an infinitely more complex object than any object researched by physics, and the human mind is even more complex (see Chapter 3 for more on this). The element of freedom and unpredictability that is inherent in the mind makes the mind a notoriously difficult object to squeeze into the full-scale replicability demand. For instance, mainstream psychology, which models itself on the physical sciences, targets the functioning of the human mind as if the mind were a biological machine or a complex computer in which information is transferred by means of the known physical forces. But even in mainstream psychology the replicability of results is not fully guaranteed and tends to decrease with time[37]. In contrast, parapsychology explores the possibility of the transfer of information between minds, or between the mind and an object, without engaging physical media (see Tables 0.3 and 4.1), which makes full replicability even less likely.

To summarise, the studies of parapsychological effects showed that supernatural events are not just science fiction; such events do exist, but they cannot be manipulated in the same way we manipulate physical events. For example, a stone thrown in liquid water will always sink, but the mind's ability of ESP or telekinesis can appear and disappear in different experiments, even when the experimental methods stay the same. *There is only a probability of previously observed parapsychological effects being replicated.* This may not be good enough for physicists and for the purpose of practical applications of these effects, but for the purpose of examining whether it is possible to observe supernatural events by means of an experiment, the reviewed effects are sufficient.

Indeed, unlike physical phenomena, *most mental phenomena are unique and unreproducible at full scale.* For example, we don't doubt that such phenomena as love, hope and belief exist. But all these phenomena are unique. We cannot love the same person twice; even less can we love a person because we are told to do so. And what about art? The paintings by Picasso, Rembrandt and Modigliani are unique and impossible to replicate with exact precision. Dreams fall into the same category. Not a single dream ever looks exactly like any other dream, yet we see dreams all the time. The magical phenomena of consciousness, studied by parapsychology, are the phenomena of the mind and not of nature. Their full-scale replication may not be guaranteed, yet they could be real. Shamans, priests, writers, artists and psychotherapists have been dealing with such phenomena for millennia, and now scientists, too, have to learn to deal with such phenomena.

The fact that paranormal effects of the mind do not conform to the known laws of nature explains why these effects are strictly 'local': they appear only in specially designed conditions, are 'subject specific' (in the same conditions, some subjects display the effects while others don't) and are not fully replicable. The effects are also time specific: they could show up at the beginning of an experiment and fade away by the end of the experiment; the same person might be able show these effects and then lose the ability entirely. When psychologists report studies on *normal psychological processes*, the results are supposed to be *stable during the time of the experiment*. For example, we want to study how many items (e.g., digits or letters) out of a group of ten items a young person can repeat back immediately after the items are shown on a computer screen for a brief moment and then deleted. Every day we invite ten participants and continue the experiment for ten days. What we get is that, although the number of recollected digits varies from person to person,

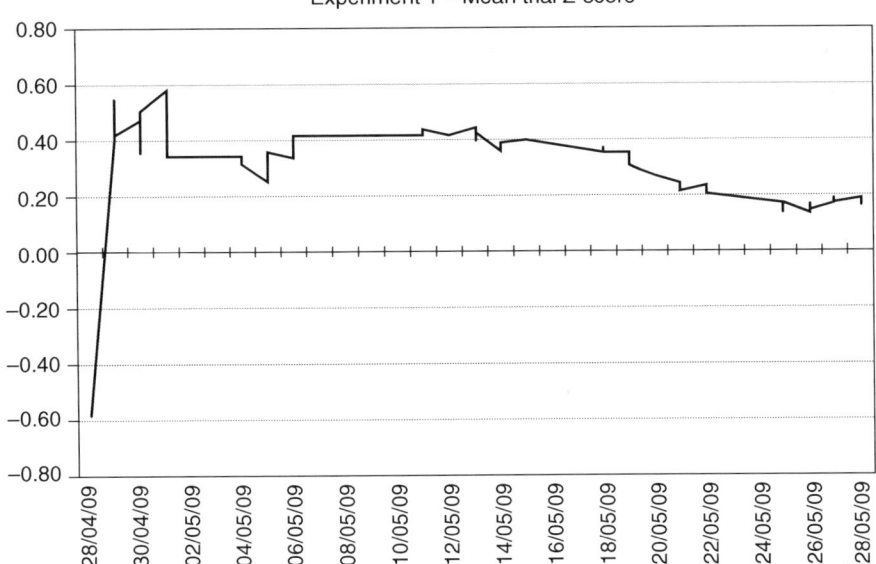

FIGURE 6.1   Decline of the paranormal effect with time of testing in Study 1, Experiment 1. Mean trial Z-score is a measure of the paranormal effect

Source: Subbotsky & Ryan, 2014

the average number that people remember on each day of testing will be about seven, and this number will stay approximately constant over the whole period of testing, given that our method and the pool of participants stay the same. In contrast, paranormal effects could significantly decline with the time of testing, even if the conditions of the experiment are maintained constant. For instance, paranormal effects significantly declined with time of testing in both of my studies reported in this section, as Figures 6.1 and 6.2 illustrate. The decline effect escapes simple explanations, such as regression to the mean, since its initial starting point is the effect, which is significant, statistically solid and unlikely to be a random deviation.

Extinction of the paranormal effect was also evident in Study 1, Experiment 2; in this experiment, along with a decline over the time of testing, the overall effect failed to reach a significant level (see Figure 6.3). How could the decline with the time of testing and the failure to replicate paranormal effects be explained?

The answer may lie in the nature of parapsychological effects. *The local character and decline of these effects with time of testing are built into the structure of the universe, guaranteeing the stability and permanence of the laws of nature.* Indeed, if magical phenomena such as direct Self over matter/mind were of a magnitude and universality comparable with that of the effects of gravitational or electromagnetic fields, then the world as we know it would collapse. People would be able to read each other's minds, thus undermining the fundamental feature of mental reality – independence of the individual's subjective experience. Spies could read classified documents,

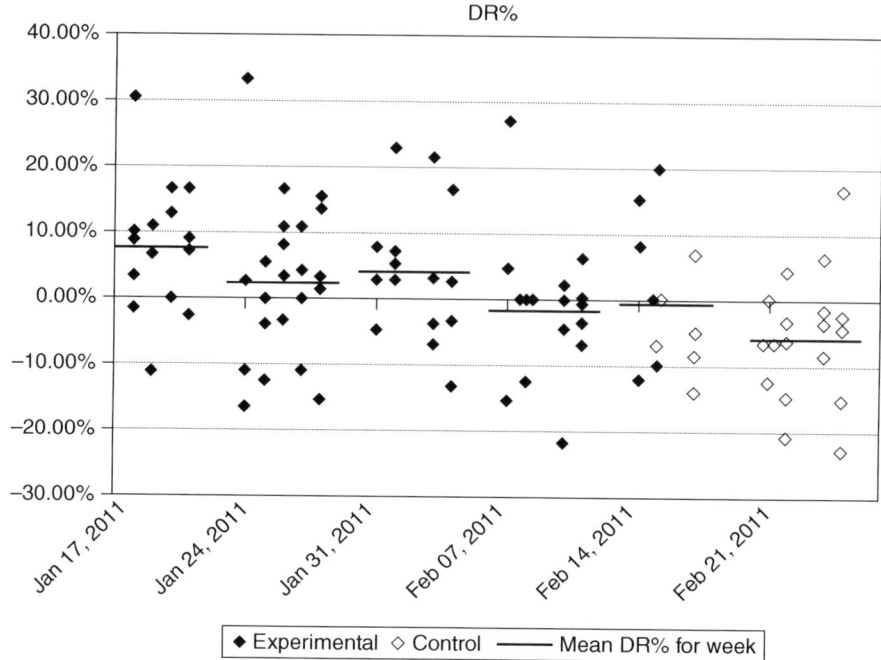

FIGURE 6.2 Decline of the paranormal effect with time of testing in Study 2, Experiment 1. 'DR%' is a measure of the paranormal effect
Source: Subbotsky, 2013

terrorists would be able to stop planes in the air just by an effort of will and people could change the orbits of planets at their whim; because human desires and goals are so diverse and controversial, the direct effects of these desires on matter would inevitably bring the universe into chaos, destroying the order and predictability of natural processes. And of course, if paranormal effects were strong, objective physical measurements would be impossible, as the measurements' results would be affected by the experimenters' hypotheses and expectations.

But if parapsychological effects are potentially destructive, why do they exist at all in the first place? Wouldn't the world be a safer place without such effects? And would it not be more logical to eliminate such effects by Occam's razor – a problem-solving principle that, when presented with competing hypothetical answers to a problem, one should select the answer that makes the fewest assumptions?

Unfortunately, this simple solution doesn't work, and the reason it doesn't work is hidden in the nature of subjective experience and in the structure of the universe. We know that for a human being the world is presented in the form of subjective experiences: we see colours, and not electromagnetic waves; we hear sounds, and not vibrations of air molecules; we taste something as sweet, not the structure of sugar molecules (see Chapter 3 for more on this). Subjective experience is a magical medium through which we perceive things-in-themselves in

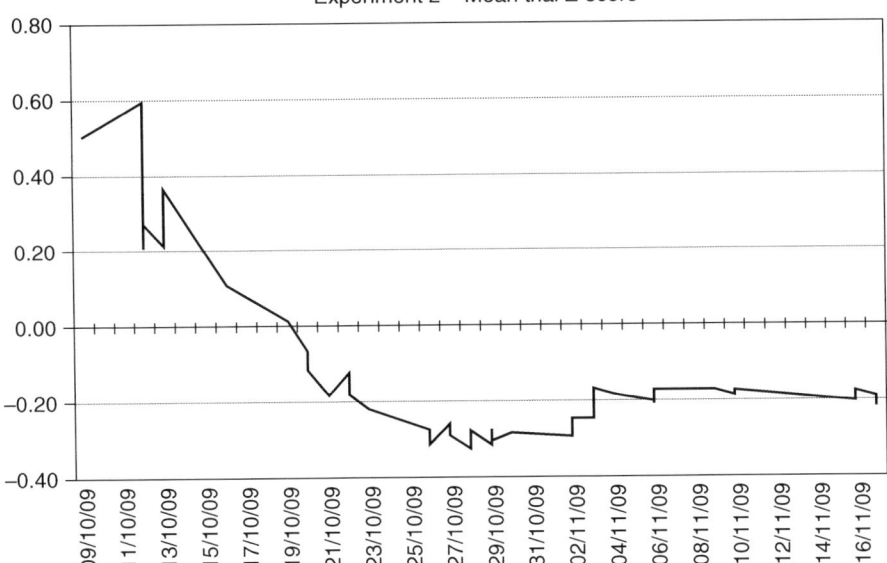

FIGURE 6.3 Decline of the paranormal effect with time of testing in Study 1, Experiment 2. Mean trial Z-score is a measure of the paranormal effect
Source: Subbotsky & Ryan, 2014

the form we perceive them. As psychological experiments show, under normal conditions, we are not even aware of this medium and confuse perceived reality with physical reality[38]. As argued in Chapter 4, although subjective experience is an extremely flexible and 'transparent' medium, like every medium it must exert a certain 'resistance' to things-in-themselves. If subjective experience didn't exert such resistance, then phenomenal images of objects would be undistinguishable from things-in-themselves, and we would have to face the absurd conclusion that *a person and the world out there are one and the same thing.* If this were the case, a person would be nothing but a soulless self-replicating machine, or indeed a puppet simulated in a cosmic computer. Another absurd conclusion that follows from the 'absolute transparency' of subjective experience assumption is that *we would not be able to voluntarily move our bodies.* Indeed, for our subjective experience (i.e., an intention to lift a hand up) to be able to affect our bodies, this subjective experience must have some ability to affect matter. Without this ability, our intentions to move would be similar to those experienced during sleep paralysis, when a person is aware but unable to move[39].

Fortunately, our thoughts and wishes do have the ability to affect matter, at least as far as it concerns our bodies and the electric activity of our brain. That is why *under special conditions the resistance of subjective reality to things-in-themselves that exist beyond our bodies has to show up, via changes of certain processes that, under normal conditions, are perceived undisturbed.* For exactly *this* reason, paranormal effects should be expected. Note that as a follow-up to this expectation, we have to accept that we *are not able to explain paranormal effects by physical causality*; there is a causal mechanism that continuously connects an act of consciousness with the changes in a physical process. Paranormal effects could only be registered as *empirical facts* of correlation between psychological and physical events. In this regard, any intentional voluntary action (i.e., an intentional lifting up of a hand) is no different from a paranormal effect (see Table 0.3).

From the previous chapters, we know that a random event is a magical event of the emergence/vanishing type that is not determined by preceding physical causes. This makes random events especially vulnerable to voluntary efforts of the mind to affect these events. In the quantum world, human subjective experience influences the probabilistic 'superposition' states of a quantum object via the measurement devices created by an observer (see Chapter 4). In the double-slit experiment *the presence of subjective experience as a medium that conducts things-in-themselves into the human mind is revealed: the quantum object, which, when its path is not detected, collapses as a wave, instead collapses as a particle when its path is known.* This effect of human consciousness on the object's behaviour is, however, masked by the presence of the observing device: it looks as though the outcome of the measurement is determined not by the observer's consciousness, but by the structure of the observing device. The presence of the observing device also makes these effects fully (or almost fully) replicable; at least, I have not come across any reports of the decline effect having been observed in the results of the double-slit experiment.

In contrast, in parapsychology, the attempt is made to affect probabilistic processes by a direct effort of the Self, which is not mediated by physical devices. As long as the Self is not a physical entity and random events do not obey physical causality, the direct effect of the Self on random processes is not a physical process either.

Paranormal effects are distortions brought into random processes by the direct effort of the Self. For example, a participant is asked to make an effort of will in order to affect an outcome of a chance process, such as the work of a random number generator based on the unpredictability of radioactive decay; as a result, one of the equally probably binary events (e.g., a flash of red light) starts to occur more frequently than the other (e.g., a flash of green light)[40]. Just like a pencil appears broken when it is half immersed underwater, certain probabilistic processes (like the work of random number generators) may become skewed if a special effort is made by a human subject with the aim of distorting these processes. These small distortions of physical processes caused by a human effort of will are what researchers call psychokinetic effects. Other 'psi' effects, such as ESP or remote viewing, are also obtained in conditions that involve random events.

To reiterate, *the magical 'psi' effects are distortions of the coordination between subjective reality and random processes at a fundamental level that occur in specially designed conditions and are not mediated by physical forces.* This may explain the decline of parapsychological effects with time of testing: when conditions are created for subjective reality to distort random processes, subjective reality quickly habituates to these conditions and has a tendency to reduce the distortions, and this explains the decline effect shown in Figures 6.1, 6.2 and 6.3. In the aforementioned physical experiments on quantum effects, this kind of habituation was not observed because observing devices mediated the link between an observer and the quantum object; the observing devices eliminated the imperfections of the observer's conscious effort, such as small deviations from the midline, tiredness and habituation, so that the habituation effect becomes hard to notice. Similarly, drawing a straight line on a computer screen with the help of an application eliminates the imperfections of the movement of our hand – the imperfections that inevitably result in deviations from a straight line when we try to draw a straight line without the application's assistance by simply moving the mouse on the table.

Because of their local character, small magnitude and tendency to decline with time, paranormal effects of subjective experience do not threaten the laws of nature. Nevertheless, these effects are important as they reveal the fact that *subjective experience is not an epiphenomenon, but a medium that possesses a certain degree of viscosity*. Thus, in chemistry, a tiny proportion of an indicator substance is needed in order to show, usually by a colour change, the presence or absence of a threshold concentration of a chemical species, such as an acid or an alkali in a solution. Similarly, *paranormal effects of the direct Self over matter type are indicators that detect the active presence of subjective experience in the universe*. This could explain why studying these effects, despite their small magnitude and incomplete replicability, evokes great interest among scientists, as well as heated discussions on the validity of these effects[41]. Much in these discussions misses the point because of the assumption that paranormal phenomena are as stable and replicable as phenomena in quantum physics. As argued in this chapter, they are not, and neither can these effects compete with physical effects on the scale of practical importance. All that these effects can prove is that subjective reality is not an illusion and can leave its mark on certain physical and psychological processes, which, under normal circumstances, are assumed to be independent of our Self. But make no mistake – we should not forget that a voluntary action of the mind on the body is a magical psi effect, too (see Tables 0.2 and 0.3), and in the grand scheme of things all of human culture and civilisation, including science, is a result of this supernatural effect.

Perhaps in the future the old principle of 'universal sympathy' (e.g., the magical unity between a person and the universe), which in the modern world has been replaced by the universal law of gravity, will eventually recover part of its former glory. As a result, humankind will cease to feel itself as being isolated and alienated in a cold and indifferent universe, and the fear of unavoidable (and possibly not-so-distant) perishing will change to a more optimistic mind-set in which the achievements of science are not opposed to magic and religion, but integrated with magical and religious forms of coping with reality.

Conclusion: the necessity of the supernatural

When Napoleon pointed out to French mathematician Pierre-Simon Laplace (1749–1827) that he had not mentioned God in his book, Laplace replied, "Sire,

I had no need of that hypothesis"[42]. In contrast to Laplace, French philosopher Voltaire (1694–1778), who was critical of official Catholic Church, nevertheless was of the opinion that "If God did not exist, it would be necessary to invent Him"[43].

We started this chapter with the following questions: is there any evidence of God's existence? Can one live without any belief at all? How do religious belief, belief in magic and belief in science come together in the minds of modern rational people? Our analysis revealed that the belief in God sprouts out of early people's belief in the magical reality of the afterlife. The spirits of the dead were the first gods. With time, the powers and responsibilities of the gods grew. From the spirits of deceased tribesmen who supervised the behaviour of the living, the gods became rulers of the elements, and eventually condensed into a 'demiurge' – a creator and keeper of the universe. The omnipotent and omniscient God appropriated most of the supernatural in the world, with humans being ousted into the world where they had to obey the merciless laws of nature: "By the sweat of your face will you eat bread until you return to the ground, for out of it you were taken. For you are dust, and to dust you shall return" (Genesis, 3:19). The fact that ordinary people could share magical powers with God came to be frowned upon. The belief that the world is essentially magical had to go underground.

However, psychological studies of recent decades have revealed that in the depth of the mind of most people, the belief in magic lives on[44]. This belief feeds into both traditional and 'rational' religious practices. This belief fuels people's enquiries into the existence of magical phenomena. These enquiries suggest that magical phenomena are at the foundation of the universe. The universe began as a magical phenomenon. Random events made it possible for the young universe to progress from simple elementary particles of matter to more complex structures and a living cell; a living cell developed into complex bodies and brains; and complex bodies and brains provided the substrate for subjective experience to interact with matter. By discovering the world of the supernatural, people developed symbolic consciousness, free will and creativity, and through these magical abilities took over the role of increasing complexity in the universe from random events. Along this line, people became aware of the supreme intelligent power that made these magical transformations possible.

Science broke into human minds like a comet and squeezed both magic and religion to the edges, but was unable to annihilate them. Science rejected magic as a fallacy. Yet it is modern physics that discovered the phenomena of magical participation between a person and the universe. Cybernetics and information theory accentuated the role of the mind in constructing our picture of the universe and creating artificial intelligence. Psychology appreciated the foundational role of human subjective experience and subconscious magical thinking in generating scientific theories. These developments revealed that magic does not contradict science; just the opposite, the existence of supernatural phenomena makes scientific enquiry more focused by outlining the limits of this enquiry. Neither should magical phenomena be viewed as undermining religion; for believers in a single god, magical phenomena should be regarded as reflections of God's powers in the facets of the crystal of the world. These reflections only point us towards the majestic source of the light.

Critical thinking makes some of us search for evidence of the existence of God. Although these enquiries are inherently contradictory (if there were evidence, there would be no need for belief), they are not meaningless. Theoretical considerations and empirical studies increasingly show that both in the human consciousness and in the outer world supernatural events do exist. The presence of supernatural events highlights the existence of an almighty power working in the universe. It is hard to express this thought better than Voltaire did in one of his letters. After his aforementioned phrase that it would be necessary to invent God, he wrote, "But all nature cries aloud that He does exist: that there is a supreme intelligence, an immense power, an admirable order, and everything teaches us our own dependence on it"[45], p. 210.

But if God exists, why is there so much suffering in the world? Why are there terminal illnesses and fatal accidents? Why do innocent people die in wars, disasters and catastrophes? The answer is – because we are alive. For living creatures, pleasure and pain, joy and sorrow, happiness and unhappiness, luck and disaster always go together. And those who live will have to die. But the distribution is unfair. Some people live long and happy lives, others are born in poverty, suffer and die young. For those who are less fortunate, philosophy and science is not much of a consolation. Only *the belief* that there is God who loves us, sees our sorrows and will put things right in life or after death *can offer a helping hand.*

Perhaps when people die they discover that there is no God. But while they are alive, the belief in God helps people to overcome suffering and gives them meaning and hope.

Notes

1 https://en.wikipedia.org/wiki/Credo_quia_absurdum
2 Dunbar, R. (2014)
3 Boyer, P. (2001)
4 Tylor, E. (1920/1871)
5 http://news.nationalgeographic.com/2015/09/150905-animals-sleep-science-dreaming-cats-brains/
6 Subbotsky, E. (2001)
7 Subbotsky, E. (2007)
8 Subbotsky, E. & Quinteros, G. (2002)
9 Davis, P. (1992)
10 Tiger, L. & McGuire, M. (2010)
11 Langer, E. J. (1975)
12 Pronin, E., Wegner, D. M., McCarthy, K. & Rodriguez, S. (2006)
13 Vyse, S. A. (2000)
14 https://en.wikipedia.org/wiki/Eucharist
15 James, W. (2012)
16 Huxley, A. (1954)
17 http://en.wikipedia.org/wiki/Marsh_Chapel_Experiment

18 Horgan, J. (2004)
19 Newberg, A. & d'Aquili, E. (1999)
20 Newberg, A., d'Aquili, E. & Rause, V. (2002)
21 Persinger, M. (1987)
22 http://en.wikipedia.org/wiki/God_helmet
23 Penfield, W. (1975)
24 Subbotsky, E. (2016)
25 Subbotsky, E. (2010a)
26 Kuhn, G., Amlani, A. A., & Rensink, R. (2008)
27 Subbotsky, E. (2010b)
28 Bem, D. J. & Honorton, C. (1994)
29 Dunne, B., Nelson, R. D., & Jahn, R. G. (1989)
30 Nelson, R. D., Bradish, G. J., Jahn, R. G., & Dunne, B. J. (1994)
31 Subbotsky, E. & Ryan, A. (2014)
32 May, E. C, Utts, J. M. & Spottiswoode, S. J. P. (1995a,b)
33 http://en.wikipedia.org/wiki/Yerkes-Dodson_law
34 Bem, D. J. (2011)
35 www.dailygrail.com/Mind-Mysteries/2014/1/Is-Precognition-Real-Positive-Replications-Daryl-Bems-Controversial-Findings
36 Subbotsky, E. (2014)
37 Lehrer, J. (2010)
38 Subbotsky, E. (1997)
39 https://en.wikipedia.org/wiki/Sleep_paralysis
40 Radin, D. I. & Nelson, R. D. (2000)
41 de Beauregard, O. C., Mattuck, R. D., Josephson, B. D. & Walker, E. H. (1980)
42 https://en.wikipedia.org/wiki/Pierre-Simon_Laplace#Quotations
43 https://en.wikipedia.org/wiki/Épître_à_l'Auteur_du_Livre_des_Trois_Imposteurs
44 Subbotsky, E. (2013)
45 http://books.google.co.uk/books?id=zmk_AQAAMAAJ

PART III

Magical thinking in politics, economics and education

7

UNDER THE SPELL

The case of Russia

Problem

In the two decades following 1991, one of the most widely discussed problems in Russia was the country's demographic situation. According to some data, from 1991 to 2012, the natural decrease of the population of Russia (not counting the increase due to the influx of migrants) was around 12.5 million (approximately 0.6 million per year)[1]. In 2013, the World Health Organization in its yearly report estimated the average lifespan of Russian men as the shortest in Europe and Central Asia – 62.8 years[2]. In his address to the Federal Body of the Russian Federation for the year 2000, Vladimir Putin said, "The number of us, the citizens of Russia, decreases from year to year. It has been for a number of years already that the country's population shrank by 750,000 with every year. And if we are to believe the prognoses, and these prognoses are based on the real efforts of people who understand the area and devoted their whole lives to the area, in 15 years from now the number of Russians may decrease by 22 million. I ask you to contemplate this figure: it is one seventh of the whole country's population. If the current tendency carries on, survival of the nation will be at stake. There is a real threat for us to become a nation of the elderly. The demographic situation today is a worrying one"[3], translated from Russian by the author].

But the demographic situation in Russia is far from being the only reason for concern. Since 1991, criminality in Russia, especially organised crime, has increased sharply. In the period of 1991–1997, the number of registered crimes doubled and the number of solved cases plummeted. With the new Criminal Code of the Russian Federation, introduced in 1997, criminality somewhat shrank, but then started to climb up again; the number of murder cases in Russia exceeded by several times those in other industrial countries, such as the USA, Great Britain, France and Germany[4]. In the new millennium, cybercrime joined traditional crime.

Various forms of corruption are also on the increase. According to *The Economist* magazine, while before 'perestroika' corruption had been qualified as a crime, after perestroika corruption became the essence of the system, with a small group of people acquiring fortunes that exceeded the wildest fantasies of the tsars[5]. According to some statistical data, in 2012 alone, 800 high-ranking officials and governors had to leave their posts due to corruption[6].

These negative tendencies, which hit Russia from the moment of the collapse of the USSR, cannot be explained by a drop in quality of life, since in the Human Development Index (a complex comparative indicator of an average individual's lifespan, literacy, educational level and quality of life in a country), published in 2012 by the UN, Russia occupied a relatively stable 55th place among 187 countries (somewhere between Kuwait and Romania)[7].

We have to acknowledge, therefore, that in the beginning of the 1990s, Russia experienced a major social and cultural shift that brought the country to the edge of demographic collapse and caused the destruction of social order and public morals. In addition, the period since 1991 brought about negative tendencies in other domains of social life in Russia, such as public health, education, science and social security. There were, of course, certain positive changes as well: freedom of speech, freedom of faith, freedom of emigration and an abundance of goods in shops and supermarkets. But these negative and positive tendencies occurred slowly and gradually, whereas the aforementioned social–cultural shift happened sharply and proved stable. This shift was not a result of external factors such as war, hunger or a sharp decline in the people's economic situation. The most astonishing feature of this social–cultural shift was that it occurred relatively peacefully, without bloodshed or any noticeable mass protests. The peaceful character of the Russian 'revolution' of 1991 looks even more surprising if compared with the revolution that happened in Russia in 1917, which was followed by a long and bloody civil war.

I leave the proper analysis of the causes of these changes in demographic and spiritual situations in modern Russia to sociologists and specialists on population growth[8]. My personal view is that one of the causes of these changes might have been the loss of the basic myth that unifies the people and creates the perspective for the people's future. Thus, in 1917, the basic myth of the Russian Empire (in the interpretation of the Deputy Minister of National Education, Count Uvarov) – 'orthodoxy, autocracy and nationality' – was replaced by the myth of 'building communism' in its Marxist version. When the latter myth was abandoned with the collapse of the USSR, accepting the new myth of 'establishing democracy and a free market economy' proved to be difficult for Russian people, because the majority of the Russian population and the existing social institutions were not yet ready to live inside this myth. As a result, Russian society plunged into a state of 'existential hollowness'. At a deep level, the people ceased to understand what kind of country they lived in and what they should strive for.

The question arises as to why the generation of Russians that grew up under the myth of communism so easily and painlessly parted with this myth. Russian sociologist Sergey Kara-Murza sees the cause of the peaceful character of dropping the myth of communism as manipulation of the mass consciousness of the Russian people by external and internal forces. He argues that for just two years (1989–1991) the ideologists of the free market economy, without giving any proof, managed to sell to workers the idea that privatisation of plants and factories and the accompanying unemployment were in the workers' own interests[9]. I agree with this argument in that the manipulation of the mass consciousness did indeed take place, yet the question remains open as to why the workers submitted to this manipulation so easily. Indeed, as Kara-Murza rightly pointed out, the manipulators had not presented any logical reason or factual evidence for their suggestion that the transition to a free market economy and unemployment were in the workers' own interests, and any reasonable person would object to the government's actions that run against the people's vital interests. So why didn't the workers object? Why did they listen to the manipulators? Why did the change to property rights and the political system in Russia, as well as the collapse of the USSR, happen so quickly and without major protests? One of the answers to this question can be found in recent psychological studies on magical thinking and magical beliefs in modern humans.

Obedience based on the belief in the supernatural

It needs to be noted from the start that this chapter is about voluntary submission of a person to the suggestions or demands of a source of authoritative influence. This source could be anyone or anything: a politician, a political party, a physician, a psychotherapist, a scientist, a financier, a charity organisation or a religious cult. It is not about obedience to demands when disobedience involves a threat to the person's life and well-being; for example, to obey or not to obey the traffic rules when you are driving a car or to stop or not to stop the car if a traffic officer commands you to stop. It is about obedience when a person has the option of not obeying without consequences. People have such an option, for instance, when they decide whether to vote or not to vote for a certain political candidate, to join or not to join a certain political demonstration, to keep taking or to stop taking an ineffective drug prescribed by their physician when the physician recommends they keep taking the drug, to invest or not to invest in a financial pyramid scheme that promises an unrealistically high profit, to buy or not to buy a brand of goods that one doesn't really need, to follow or not to follow a certain superstition when not following the superstition is believed to bring about misfortune, and so on. It might sound puzzling to a reader, but the choices we make in the aforementioned situations are determined by the ancient mechanism we keep in our subconscious – the belief in the supernatural.

The belief in the supernatural and respect for authorities

As argued in previous chapters, the belief in the supernatural (e.g., in the power of gods, spirits and wizards to produce actions that violate the known laws of physics, biology and psychology) probably emerged in the Upper Palaeolithic sometime about 30,000 years ago[10,11,12]. Being in many places simultaneously, going through solid walls unharmed, controlling the weather and crop growth, feeding on the smoke produced by burning sacrificial animals, reading people's minds – these are just a few of the vast number of supernatural powers that gods and spirits possess. When addressing gods and the spirits of the dead ancestors, people sought the gods and spirits' cooperation in solving their vital problems, such as protection from predators and success in hunting. In return, the people were happy to make sacrifices to the gods and to obey the gods' commands that they gave directly or indirectly through mediators – tribal leaders, medicine men, wizards and shamans. For instance, when asking for the healing of a sick person, people addressed a shaman; if the shaman agreed, he or she immersed himself or herself in a special meditative state, which allowed them to communicate with the gods and ask for their cooperation[13]. Recent psychological studies suggest that the belief in supernatural powers, including the belief in magic and in traditional gods, is not just ancient history, but a fundamental feature of the human mind[14,15]. Experiments have indicated that in modern industrial cultures most educated adults consciously deny that they believe in magic, yet behave as if 'at the bottom of the heart' they indeed believe in the supernatural[16,17].

In traditional cultures, to which many cultures of Africa, Asia, South America and the islands of the Pacific belong, belief in the supernatural is practiced openly. In those cultures, people take the messages of their leaders as imperatives sanctioned directly by the gods and spirits. Consequently, in these cultures, the suggestions of the leaders rarely come under scrutiny in regard to whether these suggestions are true or false. In contrast, in modern industrial cultures, people may critically analyse the messages that come from their authorities. However, in certain circumstances, even when the critical analysis shows that the authoritative message is wrong and/ or harmful to a person, the person nevertheless takes this message as an order and voluntarily follows the order. In other words, while in traditional societies people obey authoritative influences without analysing the legitimacy of these influences, in modern industrial societies people sometimes obey authoritative messages in spite of a clear realisation that such messages shouldn't be trusted and/or could bring harm to the people. Exhibited by modern rational people, obedience of this kind suggests that modern rational people subconsciously still believe in the supernatural abilities of persons of power. Psychological experiments have supported this suggestion; they have shown that under certain conditions, educated adults do believe that a person of authority (e.g., a psychology experimenter) possesses supernatural powers. One of these conditions is removing people's psychological defences against the belief in the supernatural. Such defences can be removed when an authoritative person demonstrates to participants effects that look persuasively supernatural and suggests that not believing in his or her magical powers might negatively affect the participants' future lives or their valuable objects[18,19,20].

In real life, unrealistic, objectively wrong and sometimes immoral requests can have a similar suggestive effect if these requests come from an authoritative source and target people's personally significant objects. *People's subconscious belief in the authorities' divine powers can gain the upper hand over people's critical thinking and make them obey doubtful and even obviously wrong requests.* In this chapter, this form of social obedience will be referred to as '*the belief in magic-based social compliance*' (BMSC). Let us consider three issues that this new concept raises: (a) the definitive feature of BMSC; (b) the distinction of BMSC from other forms of voluntary social submission that are not based on the belief in the supernatural; and (c) proofs of the existence of BMSC.

BMSC – the definitive feature

By definition, BMSC arises when the hidden belief in the divine powers of authorities overcomes critical thinking and elicits a reaction of submission. It is established in psychology that emotional preferences can influence reasoning, perception and memory[21,22]. For example, people may be unable to see the negative sides of a person they are in love with. The behaviour controlled by emotional preferences looks similar to BMSC. However, there is one feature that distinguishes BMSC from actions governed by our emotions: while determining our actions, our subconscious belief in the supernatural doesn't affect our critical thinking. As a result, people who act under the influence of their subconscious belief in the supernatural inevitably come into contradiction with themselves: (a) they are aware that the ideas suggested to them by the authority are wrong and run against their personal interests; and (b) they act as if the ideas were true and profitable to them.

The aforementioned contradiction is a case of word/action dissociation, when a person acts differently from what he or she promises. A typical example of this kind of dissociation is intentional or unintentional lying. For instance, in the moral domain, people can preach morality, but in their practical actions follow their selfish interests[23,24]. Individuals who act under the influence of BMSC display a dissociation that runs in the opposite direction: they might say that the offered message, demand or request is wrong and damaging to their private interests, but in reality they accept the message, demand or request as true and act accordingly. For example, experiments have shown that educated adults who verbally denied that they believed that a magic spell could influence their lives prohibited an experimenter from casting a spell in fear that the spell might actually work [18][19][20]. In this chapter, a dissociation of this kind will be referred to as an 'inverted word/action dissociation'. An inverted word/action dissociation can serve as an objective indication of the presence of BMSC. Indeed, without assuming a person's implicit belief in the authority's divine powers, it would be hard to explain the fact that the person voluntarily complies with the authority's demands that the person explicitly views as wrong and/or damaging to his or her interests. To conclude, *the inverted word/action dissociation is the definitive feature of BMSC.*

BMSC and other forms of voluntary social obedience

BMSC should be distinguished from obedience based on indirect or direct logical persuasion. For instance, the 'elaboration likelihood model' by Petty and Cacioppo[25] describes two ways to persuade people to do something. The *central way* of persuasion is based on putting forward logical proofs of why it is profitable to a person to believe a certain message and follow that message. If the person finds the logical proofs persuasive, he or she will accept the message; otherwise, the message will be rejected. In contrast, the *peripheral way* of persuasion targets the person's emotions rather than his or her logical thinking; for example, a suggestive message emphasises the authority or physical attractiveness of a messenger (e.g., the message comes from a famous scientist, an athlete, a movie star or a beautiful woman).

The feature that distinguishes BMSC from compliance based on both central and peripheral types of persuasion is exactly the inverted word/action dissociation. Compliance based on the elaboration likelihood model does not contain word/action dissociation. A person either consciously accepts the message as a persuasive one and acts accordingly or rejects the message as a wrong one and refuses to act in accord with the message. In contrast, people who act under the influence of BMSC objectively assess the message as a wrong or damaging one, yet take the message as if it were right and comply with the message. As a result, the behaviour of a person under the influence of BMSC is similar to the behaviour of a person who acts under hypnosis, with the only difference being that in the hypnotised person his or her critical thinking ability is switched off. In contrast, the person who exhibits BMSC is in full possession of his or her critical thinking abilities and acts consciously.

BMSC – empirical proofs

Like any hypothesis in science, the BMSC hypothesis can only have value if it brings us to certain nontrivial expectations that can be empirically verified. Let us consider some of these expectations.

Psychological studies have shown that when a suggested message was placed in the context of magical mythology (e.g., the messenger is presented as a wizard or a witch who uses magical incantations or a magic wand), then under certain conditions educated adults exhibited inverted word/action dissociation. For instance, participants might argue that the experimenter's assurance that his or her magic spell would affect the participants' future lives was wrong, yet act as if this message were right and so prohibit the spell [17][18][19][20]. However, *unlike in laboratory experiments, in real life, messages that appeal to people's hidden belief in the supernatural are presented to people in a mundane form and don't mention magic and witchcraft.* For example, a presidential candidate could promise, if elected, to overcome a gargantuan budget deficit without increasing taxes, or an automobile manufacturer could suggest that buying this brand of car will make a person rich. It is clear that such promises are unrealistic and appeal to people's implicit belief in the authority of the sources of these messages and not to people's logical thinking. Promises of this kind are typical of politics, commerce and pharmacology[26,27,28,29]. If the BMSC hypothesis is correct, then *a message that appeals to BMSC must produce the inverted word/action dissociation independently of whether the message is or is not explicitly associated with magic.*

With the aim of examining this expectation, an experiment was designed. The first group of participants (university graduates and undergraduates) was asked whether they would allow the experimenter to put a magic spell on their future lives (magic-loaded context). The spell was allegedly taken from an old book on magic and there were two kinds of the spell: a good spell (which promised to make the participants' future lives good and problem free) and a bad spell (which would make their lives hard and full of problems). In the interview, the participants denied that either of the spells could affect their future lives, but acted as if they really believed in the spell's magical power by allowing the experimenter to cast the good spell and prohibiting the bad one. Participants of the second group received the same instructions, but this time the instructions were free from explicit association with magic; the participants were told that if the experimenter increased or decreased the number of ones on a computer screen (e.g., changed 1111 into 11111111 or into 11), then the number of difficult problems in the participants' future lives would increase or decrease proportionally (magic-free context). The results (illustrated in Figure 7.1) were clear: although all of the participants in the second group stated that the number of ones wouldn't affect their lives, in their practical actions they behaved like the participants of the first group in that they allowed decreases in the number of ones on the computer screen but prohibited increases in the number[30]. The results confirmed the prediction: they showed that *a message that targets people's implicit belief in the authoritative person's magical powers produces the effect of inverted word/action dissociation independently of whether this message is or is not explicitly associated with magic.*

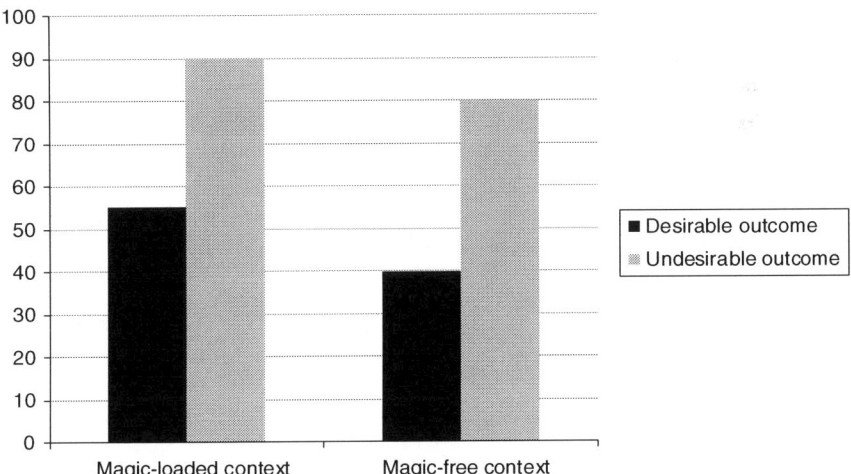

FIGURE 7.1   Percentages of participants who said 'no' to the suggested offer, as a function of the type of suggestion (magic-loaded context versus magic-free context) and type of outcome (desirable versus undesirable)

Source: Subbotsky, 2007

Another expectation that follows from the hypothesis of BMSC is that *in modern societies, social rules should exist that historically were introduced as divine and sanctioned by God*. Laws like that do indeed exist. According to the Judeo-Christian tradition, Moses obtained the moral code (the Ten Commandments) directly from God, although today most people view these laws as conventional rules developed by society. Research has shown that some individuals, both children and adults, are capable of voluntarily following these rules in the absence of surveillance, even though following the rules runs against the individuals' private interests[31,32]. This proves that implicitly these individuals view these rules as divine.

The BMSC hypothesis also implies that *a positive correlation must exist between BMSC and people's belief in the supernatural*. This means that people who are more inclined to exhibit BMSC should be more likely to acknowledge their belief in the supernatural than people who are less inclined to exhibit BMSC. Indeed, if BMSC feeds on the energy of the implicit belief in the supernatural, then in individuals whose belief in the supernatural is close to the 'surface of consciousness' and is blocked by psychological defences only to a small extent, this energy source is easier to access than in individuals whose belief in the supernatural lies deep in their subconscious and is heavily sealed by psychological defences. As a result, people of the former type will be more prone to exhibit BMSC than people of the latter type. Studies on 'interrogative suggestibility' provide experimental evidence of the positive correlation between people's tendency towards BMSC and their belief in the supernatural. *Interrogative suggestibility* is defined as the degree to which people are prone to change their opinions under the pressure of an interrogator[33]. In the experiment, a person's opinions are assessed twice on the scale of interrogative suggestibility (SIS): before and after the interrogator had tried to influence these opinions[34]. The difference between the person's initial opinions and how much his or her opinions changed due to suggestion is a version of the inverted word/action dissociation and can be interpreted as a measure of BMSC. Research has revealed a significant positive correlation between people's scores on the SIS and their scores on the scale of belief in the supernatural[35,36,37], which supports the aforementioned expectation.

Finally, BMSC helps explain phenomena that are well known in psychology but still lack a persuasive explanation. One of these phenomena is Stanley Milgram's 'obedience to authority' effect[38]. In Milgram's experiment, participants were asked to teach a 'student' (in reality the experimenter's confederate) to perform certain tasks; for every mistake, the 'teacher' was supposed to punish the 'student' with an increasingly powerful electric shock. Unbeknown to the 'teacher', the 'student' was not given the electric shock and simply simulated the pain from the negative reinforcement. The experiment's aim was to determine the magnitude of the shock (measured in volts) at which the 'teacher' would refuse to obey the experimenter's demand by ceasing to increase the voltage. Surprisingly, in one condition of the experiment, 62% of participants succumbed to the experimenter's pressure and increased the intensity of the shock up to the dangerous 450 volts; this happened in spite of the fact that a refusal to increase the shock did not involve any material or social losses for the 'teacher'. From a certain point in the experiment, the participants understood that they didn't have any moral justification to keep hurting the 'student'; the participants exhibited perspiration, their hands trembled and they began to argue with the experimenter, yet they continued to increase the voltage of the shocks. The number of participants who refused to increase the voltage of the shocks sharply dropped only when the experimenter was not personally present at the experiment but gave his commands to the 'teacher' via a telephone. As Milgram writes, "Something akin to fields of force, diminishing in effectiveness with increasing psychological distance from their source, have a controlling effect on the subject's performance" [38], p. 147. The psychological mechanisms of this 'invisible force' remain, however, unexplained.

From the theoretical perspective presented in this chapter, it is quite obvious that in Milgram's experiment participants exhibited the definitive feature of BMSC – the inverted word/action dissociation. Indeed, the participants disapproved of the experimenter's demand to keep increasing the electric shock's power and they argued with the experimenter, yet they continued to obey his instructions. In other words, the participants viewed the experimenter's demand as wrong and harmful both to the 'student' and to their own morals, yet acted as if the demand was right. From the perspective of the BMSC hypothesis, *in Milgram's experiment, the participants behaved as if subconsciously they believed that the experimenter's orders were sanctioned by supreme powers, which rendered the participants free from personal responsibility for their actions.*

One might ask what relation the participants' compliance with the experimenter's orders in the above experiment, however impressive, had with the supernatural. The answer is hidden in the history of the development of compliance with authority. Indeed, suppose that in early humans compliance with authority (a tribal leader, a shaman, a medicine man) and with tribal laws was absent. One consequence of this situation would be that each individual would have acted without coordinating his or her actions with the actions of other group members, and this would quickly end in the deaths of both the individual and the group. Early human groups did not have police forces, judicial systems and other means of law enforcement. The only way to force a person to comply with authority was to make the person believe that gods and spirits sanctioned the authority's orders (see Chapter 2 for more on that). For example, children were trained from their early years that if they did not comply with the elders' orders or the custom of sharing their food with other tribesmen, then this would infuriate the spirits of their dead ancestors, who would then punish the children for disobedience. With historic and cultural development, tribal leaders and shamans were replaced by politicians, physicians, financiers, scientists and other sources of authoritative influence, whose orders can be discussed and criticised. Nevertheless, the implicit belief that people of power have a contract with the gods can still make us obey these sources of authoritative influence, even against our will. As American psychologist Robert Cialdini puts it, in the Old Testament we read "what might be the closest biblical representation of the Milgram experiment – the respectful account of Abraham's willingness to plunge a dagger through the heart of his young son, because God, without any explanation, ordered it. We learn in this story that the correctness of an action was not adjudged by such considerations as apparent senselessness, harmfulness, injustice or usual moral standards, but by the mere command of a higher authority. Abraham's tormented ordeal was a test of obedience, and he – like Milgram's subjects, who perhaps learned an early lesson from him – passed"[39], pp. 217–218.

Altogether, the existence of BMSC confirms that, freed from explicit association with magic and renamed as 'compliance' and 'conformity', the belief in the supernatural powers of authority figures survived into the modern industrial world – the world that otherwise worships science and logical thinking.

BMSC as the economy of psychic energy

According to psychoanalysis, every conscious action requires spending psychic energy[40,41]. Psychic energy shouldn't be confused with the physical energy that we spend when we do physical exercises. Thus, after a difficult examination we feel tired like after doing hard work, despite the fact that we have not spent any physical energy. Recent neurocognitive studies have revealed that psychic energy is not a metaphor, but a real force that can be measured by cardiovascular and biochemical indicators, such as heart rate and blood glucose levels[42,43]. The experiments showed that any act of self-control decreases the blood glucose level below the optimal line, which makes subsequent actions of self-control more difficult.

Spending psychic energy occurs in any activity that includes the effort of self-control. For example, making such an effort is necessary in order to cope with worrying thoughts about the inevitability of death; as a result, thinking about death consumes psychic energy, thus decreasing the limited amount of this energy stored in the organism. In a series of experiments, one group of participants was asked to write a story about death while another group wrote a story on an emotionally neutral topic. After this, the participants of both groups were given identical tests that required a high level of self-control. The results showed that the participants who had written a story about death performed significantly worse on the tests than the participants who had written a story on a neutral topic[44].

Inhibiting the impulse to commit an action inspired by BMSC requires a substantial effort of self-control, which consumes psychic energy by taking energy from the limited reservoir. As a result, a smaller amount of psychic energy is left in order to achieve other goals that require self-control, such as political activities and achievements in education, medicine, science and art. The history of humankind shows that people usually fall under the power of dictatorships not when people prosper, but when people spend most of their psychic energy on providing for their basic needs, meaning there is not much energy left for socially creative activities. The twentieth century provides examples, such as Russia at the beginning of the century exhausted by World War I and the revolution of 1917, impoverished Italy in the 1920s and Germany in the 1930s devastated by World War I and the great economic depression. In their struggle to survive in the world of unemployment and economic devastation, in order to save their psychic energy, people invested their trust in charismatic political leaders whose orders they followed unconditionally. These examples illustrate that *the need to save psychic energy can force people to succumb to BMSC.*

In sum, recent experiments on magical thinking and magical beliefs in modern industrial cultures bring us to the following assumptions:

- The belief in the supernatural abilities of authorities is a special structure in the mind of a modern person that has its origins in prehistoric times and encourages a modern educated person to voluntarily comply with the author-ities' requests and suggestions that could run against the person's own interests, thus creating the phenomenon of BMSC.
- Science education ousts the belief in the supernatural to the domain of the subconscious and seals this belief with psychological defences.
- Depending on how deeply into the subconscious the belief in the supernat-ural is ousted, people can be divided in three categories: those whose belief in the supernatural for some reason is not ousted into the subconscious (type 1); those in whom this belief lies in the shallow layer of their subconscious and is protected by weak psychological defences (type 2); and those whose belief in the supernatural is ousted deeply into their subconscious and is sealed by strong psychological defences (type 3).
- Overcoming the tendency to act in accord with BMSC requires spending psy-chic energy taken from an individual's limited reserve of such energy, whereas succumbing to BMSC saves psychic energy.

Let's see how the aforementioned assumptions could help answer the question posed at the beginning of this chapter: why did the Russian people allow their authorities to privatise the people's property without mass protests against the obvious social injustice caused by this privatisation?

Through the magic crystal: Mexico, Russia and Europe

Any country's population is a mix of type 1–3 categories of people; neverthe-less, countries vary in regard to the distribution of people within these categories. Thus, anthropological and ethno-psychological studies of Mexican culture revealed that the belief in the supernatural is widespread among the population and openly acknowledged by its people. One of the typical folk superstitions is the belief in Nagual – a person who at night turns into an animal[45,46]. The belief in the super-natural in Mexico is particularly strong among the rural population, where this belief coexists with the Catholic faith. For example, people of the Zapotec sub-culture believe that they have two souls: a Christian soul and a 'tono' – an animal soul (e.g., the soul of a wolf or a snake)[47]. Belief in witchcraft is common even among Protestants who have a certain level of education and live in the large cities of Mexico or the USA. An experimental study of people's tendency to succumb to

BMSC conducted with uneducated inhabitants of Central Mexico (who allegedly constitute a representative sample of most of the rural population in Mexico) showed that these people predominantly belong to type 1[48]. Not only do they believe in an experimenter's magical powers, but also they openly acknowledge their belief in witchcraft. Viewed in a wider context, people of this type are not inclined to doubt the requests and suggestions of authorities. They are politically and socially inactive and are not disposed to take personal responsibility for the processes that go on in their village, city or country. In a country like that, the proportion of type 1 people reaches its maximal level. In such a country, riots and revolutions do happen, but these social movements usually have other causes than the tendency to resist the demands of local authorities; mostly, such revolts are caused by religious, national or tribal conflicts. Countries with a predominantly type 1 population are recognisable by features such as an abundance of slums, dirty and abandoned streets in the cities, poorly kept roads, poor sanitation and general shabbiness of the vital aspects of life, even in areas that do not require much investment of money and effort in order to be kept at an acceptable level. In a country of this type, it is not safe to walk the streets at night. Usually, these are poor countries with high levels of corruption and criminality and low levels of public education and medical services. Fundamental scientific studies in the country's universities are rare and poorly funded. At the same time, a country like that could strike us by the pride of its people and the beauty of its nature, art and architecture.

Type 3 people dominate in Western European cultures, such as those of the UK and Germany. This doesn't mean that the belief in the supernatural is absent in European cultures. For example, in modern England, individuals who officially practice magic are in the thousands, and these individuals come from educated, middle-class backgrounds[49]. In the middle of the twentieth century, 23% of people in England believed in ghosts, 53% had visited a fortune-teller at least once and 51% regularly examined horoscopes[50]. There is no reason to believe that today the situation is much different. Nevertheless, studies have shown that the majority of British adults consciously deny that they believe in the supernatural. Even when the psychological defences that prevent people from acknowledging their belief in the supernatural were partly removed, participants kept denying their implicit belief in magic. However, when their defences were completely removed, British participants' behaviour was not significantly different from the behaviour of uneducated adults from Mexico [18][19][20]. Nevertheless, in their everyday lives, people from countries with a predominance of type 3 citizens are in sharp contrast to people from countries with a predominance of type 1 citizens. In the former, people are significantly less reliant on the power of authorities at all levels, from a village to a parliament, than in the latter. At the same time, these populations are more patriotic and feel personal responsibility for what's going on in their village, city or country. In these countries, public places (e.g., a national reserve, a park, a public toilet) are kept in order even when surveillance is absent, and it is safe to walk the streets at night. Governmental decisions are widely discussed in the media and can be argued with. Fundamental studies in science are encouraged and supported by governmental bodies. There is a certain degree of corruption and criminality, but these social sores are under control; most crimes are solved and corrupt politicians are persecuted. Altogether, the life of an average citizen in countries with a predominance of type 3 citizens is more orderly and civilised than in countries with a predominance of type 1 citizens.

In the USSR, belief in the supernatural was generally condemned. In modern Russia, the number of people who declare themselves believers in the supernatural has grown immensely. According to statistical surveys, only 10.3% of respondents answered that they did not believe in the supernatural of some kind, religious beliefs included[51]. Experiments revealed that educated Russian participants believed to a significantly greater extent than German participants that a psychology experimenter who demonstrated an effect that looked like a case of real supernatural magic did indeed possess supernatural abilities. Russians also exhibited a significantly stronger belief than Germans in anomalous phenomena, such as the Loch Ness monster, UFOs and the yeti[52]. Altogether, as much as the behaviour of a limited sample of participants can represent a whole population, educated Russian adults (and in the USSR, most of the adult population had at least a secondary school education) fall into the type 2 category. Their belief in the supernatural is ousted into the shallow level of their subconscious and the psychological defences that seal this belief are relatively weak. In a country with a predominance of type 2 citizens, most people have a secondary school education; consciously, these people cherish science and are capable of critically assessing the suggestions and demands of authorities. At the same time, the average citizen is inclined to load the burden of taking important social decisions on the authority's shoulders and does not feel personal responsibility for such decisions. While being critical of decisions made by governmental institutions, the citizen nevertheless voluntarily complies with these decisions. In their everyday lives, people from countries with a predominance of type 2 citizens stand in between countries with a predominance of type 1 or type 3 citizens. Whereas the cities are kept in order, it is not always safe to walk the streets at night. Public places that are not under surveillance (e.g., a meadow in a forest, a public toilet) are not always clean. Corruption and criminality are at high levels, most crimes remain unsolved and punishments for corruption are absent or symbolic. Patriotic feelings are low and small businesses are depressed by corrupt local authorities or are controlled by organised crime.

It would appear that Russia has all the necessary conditions to become a prosperous European country: huge natural resources, an educated population, a parliament, a democratic constitution and freedoms of speech, faith and movement. However, the country's development is at a standstill. Tremendous financial flows from Russia go into the banks of foreign countries, the production of goods (except arms production) is at a low level and financial support for fundamental scientific research often doesn't reach its destination and dissolves within bureaucratic structures, forcing young and talented Russian scientists to search for work abroad. There is an impression that the country, with such huge cultural and productive potential, is under a magic spell, which it is desperately trying but failing to shake off.

Who put the spell on Russia? The historical roots of BMSC

The origins of a nation's BMSC type are rooted in the nation's history. Thus, in the peoples of the Aztec Empire, the belief in magic was a part of the main religion. With the conquest of Mexico by the Spanish in 1512–1521, the indigenous population was converted to Catholicism. However, in Mexico, Catholicism was tolerant of traditional pagan beliefs in the supernatural. Persecution of witches by the Holy Inquisition in Mexico did not cross the borders of Mexico City and did not involve the indigenous population. For the whole period of the Spanish Inquisition in Mexico (1571–1820), only around 50 people were executed, and only some of them were put to death for witchcraft[53]. As a result, the majority of the Mexican population stuck to their traditional beliefs in magic, which is currently mixed with Catholic beliefs in Christ and the Virgin Mary.

In Western Europe, in the time of Roman Empire, the belief in magic was a part of official religion as well. With the adoption of Christianity by the Roman Empire in the fourth century CE and during the Dark Ages, the belief in magic continued unperturbed. However, in 1484, Pope Innocent VIII issued a papal bull that condemned witches in Germany. The Holy Inquisition was founded and the witch-hunts began. According to some estimates, during the whole period of the witch-hunts in the Western world, up to 200,000 persons were tortured and executed[54]. As a result, an average inhabitant of Western Europe developed *a fear* of being accused of witchcraft, which ousted the belief in magic to the subconscious. This fear became one component of the psychological defence that prevents the belief in magic from entering the domain of conscious awareness. The official belief in Christ and Scripture became the only area where miracles were allowed to happen. With the emergence of science in the sixteenth and seventeenth centuries, the *shame* of believing in the supernatural joined the fear of being accused of witchcraft. Science condemned the belief in the supernatural as a fallacy, and medicine proclaimed this belief a clinical disorder. As a result, an average European citizen developed BMSC of the third type, in which the belief in the supernatural is ousted deep into subconscious and is sealed by psychological defences of fear and shame.

Before the onset of Christianity at the end of the tenth century, magic in Russia was a part of pagan polytheistic beliefs. Like the Catholic Church, the Russian Orthodox Church condemns magic for its link with the devil. Nevertheless, in Russia, there was no Inquisition, and persecution of witches was significantly less intense than in Western Europe. The number of executions of alleged witches from the eleventh until the nineteenth centuries is estimated to be only in the tens (perhaps hundreds)[55]. As a result of reforms initiated by Peter the Great in the first half of the eighteenth century, there appeared educational institutions of the European type, and science made its way into Russian culture. In the Soviet period, official communist ideology suppressed beliefs in God and magic, but the pressure on magic from religion became less intense. As a result, in the Russian version of BMSC, the belief in the supernatural was ousted into the subconscious not as deeply as in Western European inhabitants, and the psychological defences against the belief in the supernatural in the Russian people – fear and shame – were far not as strong as in Western Europeans. This resulted in the evolvement of the intermediate, second type of BMSC, which is characteristic of most of the population in modern Russia.

The peaceful revolution

In modern Russia, ideologists of 'perestroika' skilfully used the Russian people's inclination toward BMSC in order to privatise public property. At the end of the 1980s, when the Soviet people unconditionally trusted their government and official mass media, the central channels of Soviet TV were given to the 'wizards' – magical healer Allan Chumak and hypnotiser Anatoly Kashpirovsky. Millions of people drank water magically 'charged' by Chumak or moved their heads while sitting in front of the TV on the orders of Kashpirovsky. By the law of big numbers, some people indeed felt as if they experienced an easing of their sufferings from these actions due to hypnosis and the placebo effect. In the beginning of the 1990s, the 'wizards' suddenly disappeared from the TV screens and were replaced by the ideologists of the free market economy. The new showmen put forward the idea of 'privatisation', but in reality, *planned to strip the people off their property and move that property into the hands of the ruling party and KGB authorities, their friends and their relatives.* Although every person with a certain level of education (and those constituted the majority of the Russian population) clearly understood that privatisation of communal property would lead to drastic inequalities, the deal was done. The 'wizards' strengthened the people's subconscious belief that the authorities had the undisputable divine right to decide what to do, and the people accepted the idea of privatisation without resistance.

Unlike Stalin, who was adored by the Russian people, Boris Yeltsin was not viewed as a living god – his human weaknesses and passion for drinking were widely known and openly ridiculed by the population. And yet Yeltsin managed to do what seemed to be impossible – ending the Soviet Union, placing a ban on the Communist Party of the USSR and taking communal property from the people's hands. Like the Russian tsars, who passed their power to their heirs, Yeltsin passed his 'magic rod' of power to Vladimir Putin. That is why Putin – a mid-level officer in the Soviet KGB, unknown to the people, not a leader of a political party and without original political ideas of his own – suddenly became the victor in the presidential campaigns: succumbing to BMSC, the majority of the Russian population voted for Putin voluntarily when the holder of magical power – Boris Yeltsin – said the people should do this.

What to do?

So, how could the magic spell that fell on Russia be shaken off? What could Russian people do in order to free themselves from the inclination of complying with the orders of authorities when those orders run against the people's conscious intentions? And what should they strive for now, when the deal is done and the people are in the middle of the trap? Do they need a new social revolution? But history shows that social revolutions do not free people from BMSC. Englishmen executed Charles I of England, but got Charles II instead. Frenchmen executed Louis XVI of France, but got in exchange the Great Terror of the French Revolution and then Napoleon, who wasted 1.5 million of the best Frenchmen in incessant wars. Russians executed Nicolas II and got in exchange Lenin, Trotsky, Stalin and the Red Terror as well. So, another social revolution doesn't seem like a good choice.

Rather, what Russian people need to strive for is what the Spanish philosopher José Ortega y Gasset called "the revolt of the masses"[56]. By the revolt of the masses, which happened in the first half of the twentieth century, Gasset meant not armed struggle, but relatively fast (on the scale of history) alterations of the masses' psychology in Western Europe. As a result of these alterations, the world has drastically changed. While before the year 1920 the masses in Europe patiently endured a modest existence in the shadow of culture, in the following three decades they demanded their fair share of cultural and material wealth: a proper salary, a family car and the opportunity to educate their children, to spend their holidays in nice resorts and to attend theatres, libraries and restaurants. In other words, the places that traditionally only the rich and famous could afford suddenly became rather crowded. If the 'revolt of the masses' notion is projected onto the problem of freeing the Russian people from the tyranny of BMSC, it becomes clear what the Russian people should strive for. They should strive for the situation in which an average Russian citizen ceases to view the authorities' suggestions as if God sanctioned them; simultaneously, the citizen should stop shifting the responsibility for his or her life conditions onto the authorities' shoulders. What one needs is to change the feeling 'they are deceiving me' to the feeling 'I am allowing them to deceive me'. The understanding should appear that order in the country starts with order in the street next to your house, with a clean backyard, with not leaving garbage in the forest meadow after your picnic, with your tolerant and decent behaviour on public transport, with not passing by when you see that an innocent person is being hurt and a thousand of other small issues that our everyday lives are made from. Understanding this and putting such an understanding into action would indeed be the revolt of the masses, which, together with other positive changes in the country, could take the magic spell off Russia and turn Russia from a type 2 into a type 3 country. Let us try to analyse what could be done in order to assist this process and liberate the Russian people from the power of BMSC. Since we already know the psychological causes of BMSC, we can consider various scenarios that might lead to such a liberation.

The most simple and easy way is going 'top down' – supplying the population with the psychic energy necessary to resist the tendency to BMSC. For example, the population can be provided with good salaries and shorter working hours. This was the approach the USA took in order to change the economic and psychological situation in post-war Germany and other European countries whose people suffered as a result of World War II. According to the so-called 'European Recovery Program' (sometimes also called 'the Marshall Plan'), in 1948, West Germany and other European countries received $13 billion (approximately $132 billion in current dollar value as of October 2017) in support to help rebuild their economies. The USSR refused to receive this kind of help, and modern Russia is unlikely to get it either; rather, Russia is more likely to suffer more economic sanctions. And even if financial help from abroad were given to Russia, it is unlikely to reach the desired destination and would dissipate into corrupt bureaucratic structures.

The opportunity remains, though, to go the 'bottom up' way, relaxing the power of BMSC by decreasing the population's subconscious belief in the authorities' supernatural powers. As history has shown, authorities encourage this subconscious belief in order to strengthen their grip over people. Usually, they do this by displaying miracles. For example, the Biblical character Aaron magically turned his walking stick into a serpent (Exodus, 7:10). In the New Testament, Jesus expelled demons from a madman, walked on the water, fed 5,000 people with five loaves of bread and two fish (Matthew, 14:13), resurrected a dead mad and was himself resurrected.

The main problem, therefore, is to find a way to weaken the tendency towards BMSC and thus to liberate the masses from compulsively compliant behaviour, making them capable of relying on their own critical thinking instead. The approach of pushing the belief in the supernatural deeper down into the subconscious, which historically Western Europe has gone through, is impossible for Russia, not only because witch-hunting and burning people at the stake are things of the past, but also because this approach requires conditions that are absent in modern Russia – centuries of time and a fanatical religious zeal in the population.

A more realistic approach would be to weaken the source of the BMSC energy – the fear of the supernatural. How could this be done? We know that in modern educated adults their belief in the supernatural resides in the subconscious, so a direct onslaught with the aim of undermining this belief is unlikely to be effective. Besides, proclaiming the belief in the supernatural 'a fallacy' would be a lie, since there is truth in this belief[57]. Fortunately, what needs to be targeted in this particular case is not the belief in the supernatural as such, but a certain kind of this belief – the belief that the demands and suggestions of authorities are sanctioned by God. To do this, a method developed by psychoanalysis could help: making a patient aware of the fact that he or she lives under the spell of a certain hidden experience – 'the complex'. The result of this 'assisted awareness' might lead to the dissipation of the 'complex' (in our case, the people's tendency to unconditionally comply with the demands of authorities). In order to achieve such awareness, we need to relax the psychological defences – fear and shame – that prevent the belief in the divine power of authorities from becoming conscious. These psychological defences can be relaxed through a set of educational programmes on magical beliefs in modern people.

First and foremost, *a widespread sentiment that magic is a dark force (the opinion of the Russian Orthodox Church) or a fallacy equivalent to clinical pathology (the opinion of official science and medicine) should be overcome.* Equally, inadequate public images of magic are created by appearances on TV and in media of various 'witches', 'psychics' and other charlatans who exploit the people's subconscious belief in the supernatural for financial and social gain. *In contrast to these distorted images of magic, public education should explain that magical thinking and behaviour are inherent features of the human mind.* In the logical plane, *magic is not an enemy, but 'a black swan' of science.* Each time we think about science, we – consciously or not – at the same time have to think about a concept opposite to science, and that concept is magic. In the historical plane, magical thinking gave birth to science and continues to feed science with creative ideas. Alchemy gave birth to chemistry, astrology preceded astronomy and the magic of numbers was at the foundation of modern mathematics. The gods that roamed the sky on chariots became modern planes and spacecrafts, and the wizards that could speak to people remotely turned into modern radio and television. Similarly, *the belief in the supernatural is not an opponent of modern religion, but an important component of the belief in the Almighty God.* Without the beliefs of an ordinary person that he or she could communicate with God via payer, that God is somehow supervising the person and that God loves the person and is ready to help, modern religion would be dead.

Second, *the positive functions that magical thinking and beliefs play in everyday life should become widely known, as well as the ways in which magical thinking and beliefs can be misused by various organisations and individuals.* While magical phenomena are rare in the physical world, they abound in our psychological world. A magic spell cannot move a rock, but it can change our thoughts and feelings. A spell can elicit fear or hope, and these feelings can push us into action. In other words, much in our mental world obeys the laws of magic (see the previous chapters for more on this). Studies have shown that the belief in the supernatural could give a person the feeling of control over some events in his or her life[58]. Engagement with magical thinking could enhance divergent creative thinking in children[59], increase children's ability to distinguish fantasy from reality[60] and help viewers better remember advertised products[61]. Discussing in the media some of the unusual features of human psychology, such as the placebo effect[62,63] and the paranormal phenomena of the mind[64,65], could also help. When in the Russian culture *the understanding becomes established that the magical impacts of a symbol, image and poetry on the human mind and behaviour are not just metaphors but realities, then the fear of magic as something akin to a dark force will fade away.*

At the same time, people need to become aware of the fact that, along with their positive uses, magical beliefs could also be misused to manipulate the mass consciousness. Not only leaders of controversial religious cults, commercial advertising and charlatans of all sorts who present themselves as healers and fortune-tellers, but people's own democratically elected governments and presidents could use people's subconscious magical beliefs with the aim of extracting social, political and financial gains. *The people should realise that their subconscious belief in the divine power of authorities undermines their critical thinking and weakens their ability to consciously participate in the political life of their country. This awareness could diminish people's tendency towards BMSC and make them more self-conscious, responsible and politically active citizens.*

The aforementioned educational efforts alone *could not drastically change the Russian people's 'national character' and their adherence to BMSC.* For such a change to occur, steady and slow improvements in all aspects of life – economic, social, juridical and political – are needed. It is difficult to tell how long such improvements might take to come about, but educational efforts could certainly contribute to this change.

Conclusion: moving forward will be hard, but not impossible

So, the analysis brings us to a somewhat disappointing but not completely hopeless conclusion: diminishing the grip of BMSC requires steady educational efforts in the media, improvement of the general well-being of the population and a more effective struggle with corruption and organised crime. A one-step jump from type 2 to type 3 BMSC is hardly possible. Yet it is important that people know about the subconscious belief in the supernatural and how this belief gives rise to BMSC. To know this is important in order to understand that the most difficult obstacle to Russia becoming a prosperous democratic country is not the intrigues of internal or external enemies, but is hidden inside the minds and hearts of the Russian people.

Notes

1 http://ru.wikipedia.org/wiki/Население_России
2 http://argumenti.ru/society/2013/03/239412
3 www.intelros.ru/2007/01/17/poslanie_prezidenta_rossii_vladimira_putina_federalnomu_sobraniju_rf_2000_god.html
4 Dolgova, A. I. (2001)
5 www.economist.com/node/21548941
6 http://ru.wikipedia.org/wiki/Коррупция_в_России#cite_note-Econ_2012-72
7 https://data.undp.org/dataset/Table-2-Human-Development-Index-trends/efc4-gjvq
8 www.modern-rf.ru/otsenka-situatsii/otsenka-situatsii_59.html
9 Kara-Murza, S. G. (2007)
10 Ingold, T. (1992)
11 Mithen, S. (2005)
12 Whitley, D. S. (2008)
13 Lévy-Brühl, L. (1985/1926)
14 Boyer, P. (1994)
15 Nemeroff, C. & Rozin, P. (2000)
16 Subbotsky, E. (2010b)
17 Subbotsky, E (2014)
18 Subbotsky, E. (2004)
19 Subbotsky, E. (2007)
20 Subbotsky, E. (2009)
21 Forgas, J. P. (2002)
22 Gasper, K. (2004)
23 Batson, C. D. & Thompson, E. R. (2001)
24 Wicker, A. W. (1969)

25 Petty R. E. & Cacioppo J. T. (1986)
26 Castiglioni, A. (1946)
27 Coriat, I. H. (1923)
28 Malinowski, B. (1935)
29 Tambiah, S. J. (1990)
30 Subbotsky, E. (2007)
31 Subbotsky, E. V. (1983)
32 Batson, C. D. & Thompson, E. R. (2001)
33 Gudjonsson, G. H. (1987)
34 Gudjonsson, G. H. (1984)
35 Haraldsson, E. (1985)
36 Hergovich, A. (2003)
37 Petsa, E. (2012)
38 Milgram, S. (1992)
39 Cialdini, R. B. (2007)
40 Freud, S. (1935)
41 Jung, C. G. (1960)
42 Benton, D., Parker, P. Y. & Donohoe, R. T. (1996)
43 Fairclough, S. H. & Houston, K. (2004)
44 Gailliot, M.T., Baumeister, R. F. & Schmeichel, B. J. (2006)
45 Sejourne, L. (1976)
46 Redfield, R. (1968)
47 Selby, H. A. (1974)
48 Subbotsky, E. & Quinteros, G. (2002)
49 Luhrman, T. M. (1989)
50 Gorer, G. (1955)
51 www.gumer.info/bibliotek_Buks/Sociolog/Article/mchedl_vera.php
52 Subbotsky, E. & Trommsdorff, G. (1992)
53 http://en.wikipedia.org/wiki/Mexican_Inquisition
54 https://en.wikipedia.org/wiki/Witch-hunt
55 http://vk.com/topic-55746386_28497423
56 Ortega y Gasset, J. (1932)
57 Subbotsky, E. (2010b)
58 Langer, E. J. (1975)
59 Subbotsky, E., Hysted, C. & Jones, N. (2009)
60 Subbotsky, E. & Slater, E. (2011)
61 Subbotsky, E. & Matthews, J. (2011)
62 Hróbjartsson, A. & Norup, M. (2003)
63 Kaptchuk, T. J., Friedlander, E., Kelley, J. M., et al. (2010)
64 Bem, D. (2011)
65 Subbotsky, E. (2013)

8

WATCHING THE IMPOSSIBLE

Educational effects of magical thinking

Problem

Although the world of children today is full of magic and fantasy (e.g., fairy tales, cartoons, books and computer games in which the laws of science are frequently violated), relatively little in known of the role that magical thinking plays in children's development and education. Thus, we know that even before children learn to read they like listening to fairy tales and watching cartoons with magical events. For example, ever since my son was 1.5 years old, he enthusiastically watched the marvellous cartoons with the magical character Moonzy (a creature from the moon). I have to confess that my wife and myself enjoyed watching these cartoons, too. For more advanced ages, there exist such masterpieces of literary magic as the fairy tales by the Brothers Grimm and Hans Christian Anderson, *The Chronicles of Narnia* by C.S. Lewis, *The Lord of the Rings* by J.R.R. Tolkien, J.K. Rowling's *Harry Potter* series and the peak of magical fantasy – *Alice's Adventures in Wonderland* by L. Carroll. And what about children's role play? Games of pretend abound with magical events. In play, a Lego block can turn into a piece of cheese and a wooden stick into a horse. When engaged in a game of pretend, children can fight monsters and dragons, speak with trees and breathe underwater like a fish. Most adults encourage children's fantasy games and beliefs in Santa and fairies by pretending that magical events and characters really do exist. The onset of the gadget world in the twenty-first century made access to a magical Wonderland easier still. In the digital realms of Minecraft or Roblox, children can create whole new worlds by touching the screen of an iPhone with their fingers.

Regarding the role of fairy tales in children's psychological development, psychologists' views vary. Thus, American psychologist Bruno Bettelheim argues that fairy tales help children understand the inner conflicts that they experience in their spiritual and intellectual development and resolve these conflicts in a symbolic plane. Fairy tales also enrich children's knowledge of life and help them cope with fears of the horrible[1]. At the same time, some psychologists have suggested that children's early representations of the world could be an obstacle to learning science concepts later in life. For example, four-year-olds' naïve belief that all things in nature exist in order to serve people's needs (e.g., 'a lion is for a zoo', 'a cloud is for rain') provides fertile soil for these children's later beliefs in creationism (the belief that animals and people were created by God) and resistance to the scientific theory of evolution through natural selection[2]. Extending this argument, one might think that fairy tales and games of pretend, which incorporate magical events, draw children away from reality and inhibit their ability to absorb the concepts of science.

A recent neurological study has shown that listening to an extract from the *Harry Potter* series that described supernatural magical effects was associated with a stronger activation in certain parts of the brain (such as the left amygdala) than listening to a similar text without supernatural events[3]. Retrospectively, this gave support to the suggestion that exposing people to supernatural versus non-supernatural contexts could have different effects on cognitive functioning, including learning. The problem is finding out which of the two types of information – ordinary or supernatural – facilitates learning to a larger extent. The existing evidence on the effect of fantasy on learning is mixed. Some evidence suggests that framing cognitive tasks within a fantasy context facilitates thinking. For example, studies have shown that embedding a logical task within a make-believe context improved four- and six-year-old children's ability to make correct logical inferences from counterfactual premises, as compared with tasks presented within a matter-of-fact context[4]. Other research has confirmed the facilitative effect of fantasy contexts on children's performance on cognitive tasks[5,6,7,8,9,10,11]. However, there is also evidence suggesting that a fantasy context can be less favourable for cognitive functioning than a real-life context. Thus, research has revealed that preschool and elementary school children were less likely to transfer solutions from stories about fantasy characters to real-life tasks than from stories about real people to real-life tasks[12,13].

Altogether, the reviewed studies raise the following problem: keeping in mind children's psychological development, how should we treat young children's fascination with magical things? *Should we encourage these beliefs as something that benefits children's emotional and cognitive development or should we disapprove of these beliefs because they might impede children's ability to learn the concepts of science?*

In order to clarify the problem, let us break it down into three simpler problems. First, it is necessary to answer the following question: *do preschool children really believe in magic or do they simply pretend that magical events and characters are there without really believing in them?* Clearly, if children believe that the magical events they see in movies or read about in books can happen in real life, then the impacts these events have on children's thinking, perception and memory are likely to be greater than if the children think that these magical events are just fantasies. This assumption has been partially supported by experiments. These experiments showed that when children believed that a TV clip in which characters behaved badly was made up, they were less likely to imitate the characters' violent actions then if the children thought that the clip was a documentary, portraying events that had really happened[14,15]. By projecting these results onto the effect of magical beliefs on science education, we can assume that if children don't believe in the reality of the magical events they play out, then playing out these events cannot possibly do any harm to their subsequent science education. If, however, the children believe that these magical events are real, then the issue about the inhibiting effect of the belief in magic on children's scientific education deserves a more serious examination.

The second question to ask is: *what kinds of books and movies elicit a stronger interest in children: books and movies with fantastical magical events or books and movies with novel and interesting physical events?* If books and movies with fantastical magical events are less interesting to children than books and movies with novel and interesting physical events, then children's engagement with magical content cannot be a cognitive impediment to learning science. The reason for this is that books and movies with magical events simply won't be able to compete for the children's interest against books and movies that contain fascinating scientific effects. If, however, books and movies with magical events prove to be more interesting than books and movies that depict scientific events, then the children's interest in magic may indeed get in the way of children's science education.

Yet even if the answers to the aforementioned questions show that children believe in magic and are more interested in magic than in science, a third question can be asked: *instead of viewing children's fascination with magic as an obstacle to science education, couldn't this fascination be used to boost their learning about science?* Indeed, magical events, by definition, are reflections of scientific events in the 'mirror of the supernatural'. While watching magical events or reading about such events, children might ask themselves the following questions: "Why can't a person become invisible in the real world?" "What laws of nature prevent me from flying on a broomstick in my back garden?" As a result, watching a movie with magical events, instead of diverting children from thinking about science, could make them reflect upon the laws of nature that render magical events impossible in the real world. Perhaps engagement with magic could also have other, more specific positive effects on children's thinking, perception and memory.

Let us try to answer the aforementioned questions one by one, and then return to the main question asked in this paper: *is children's early belief in magic an impediment to science education?*

Magical thinking and belief in magic in children

As argued in the Introduction, scientific thinking follows the laws of formal logic and is based on the assumption that the world is built on the universal and unchangeable laws of nature. By contrast, magical thinking allows for instances where these laws are violated. When watching movies or reading stories with magical events, we inevitably and unwittingly engage in magical thinking. Because magical thinking unfolds in the domain of the imagination, it does not necessarily challenge the belief in that the real world strictly conforms to the laws discovered by science.

Nevertheless, as shown in the previous chapters, some people consciously believe that supernatural things can happen in the real world. This divides people in two categories: *those who imagine the supernatural without really believing in it and those who believe that supernatural phenomena exist not only in their imaginations but can in fact be real.* The question therefore is to find out which of these two categories preschool children belong to when they play with the supernatural. Do they believe in magic or do they only play with magical events and characters without really expecting these events and characters to exist in the physical world?

Studies on children's thinking revealed that in their reasoning children often exhibit a belief in the supernatural. Jean Piaget reported that Swiss children at the age of four to seven years often attributed the abilities of thinking and feelings to inanimate objects[16]. For instance, the children believed that a twisted rope 'wants' to unwind itself because 'it knows that it is twisted'. The children also thought that their magical actions could have a direct effect on the real world. Thus, one child said that if he sacrificed his favourite toy, then his sick mother would get better. We should bear in mind that Piaget conducted his interviews with children around 90 years ago and today children of the same ages might answer his questions differently. Another limitation of Piaget's studies was that he only studied children's verbal judgements about magic, without testing whether the children's verbal magical beliefs had any impact on their behaviour. It is therefore necessary to examine whether modern children, like those tested by Piaget, are prone to magical thinking, and if they are, then whether they really believe in magic.

In order to answer these questions, four-, five- and six-year-old children in Moscow were told a story of a magic box that could turn pictures of objects into real versions of the objects shown on the pictures if three magic words were spoken. Most children denied that this was possible in the real world. However, when the experimenter left the children alone in the room, 90% of them tried to convert pictures into objects and were very disappointed when their magic words didn't work[17]. In another experiment, children of the same age groups were told a story of a girl who was given a small table that had the magical ability to turn toy animals into real ones if the toys were put on the table. Again, when the children were asked whether a table like that could exist in real life, only one four-year-old said 'yes'. However, when the children were left alone in the room that contained a small table and toy animals, a great majority tried to put toys on the table. The table and one of the toys had magnets hidden inside. When a toy lion started to move on the table as if by itself, only a few of the children behaved as 'little scientists' would do: trying to find the mechanism behind the movement or looking for the wires connecting the table to the mains. The majority of the children, however, either left the room in fear that the lion was really coming to life or used the 'magic wand' they had been given in order to stop the conversion process. A similar belief of preschoolers in magic was also found in British[18] and American[19,20] children. Studies revealed that under certain conditions young children could even be made to believe in a novel fantastical entity they had never heard of before[21]. These studies have shown that by the age of four years, most modern children have learned that the supernatural can only exist in fairy tales. This knowledge, however, remains only theoretical; in practice, most four-year-old and even six-year-old children behave as if they believe that the supernatural could happen in real life.

Altogether, *the studies confirm that modern preschoolers not only play games of pretend with magical content, but also believe in magic.* This brings us to a conclusion that children's belief in magic can indeed create a competitive alternative to their belief in science. But is the children's interest in magic really that strong? Is this interest comparable with their interest in unknown physical events?

Magic versus physics: what is more interesting?

It is common knowledge that human and animal learning can be based on a variety of motives. For example, psychologists distinguish between external and internal motivation. External motivation is grounded in our basic needs, such as the need for food or self-preservation. One more external motive could be a child's need for positive relationships with close adults. Finally, a child's desire to play can be used as an external motivation for learning as well. For example, children may be asked to do their homework first and then be given a gadget on which to play their favourite games.

A more effective way to motivate children to learn is to make the learning interesting in its own right. Studies have shown that internal motivation for learning positively correlates with both children's scores on special tests and their marks at school. By contrast, external motivation negatively correlates with school achievements[22]. In other words, the more interesting a certain learning content is for a child, the easier it is for the child to learn and remember this content. The question is, *what kinds of events make for more interesting content for children: supernatural events or equally interesting and novel natural events?* What attracts children more: magic or physics?

For such a comparison to be valid, one has to ensure that the supernatural and the natural events are equally novel to the children. Indeed, one of the factors that makes a subject interesting for a child is the subject's novelty[23,24]. However, novelty alone is not sufficient to elicit exploratory behaviour in children; in order for something to become an interesting object for exploration, it has to not only be novel, but also attractive. Thus, novel foods do not elicit in people a desire to taste them if these foods look unattractive or dangerous for consumption[25,26]. The feature that makes a new phenomenon attractive to explore is the phenomenon's subjective value to the person (i.e., their capacity to satisfy the person's basic needs). One might assume that *observing a supernatural event is subjectively more valuable to a person than observing a novel natural event.*

Indeed, the ability to change objects with the help of magic words makes a child an incomparably more powerful person than the ability to change the same objects with the help of physical devices. Magic opens up to the child an opportunity to become a powerful person here and now, whereas science demands years of hard work and training. It is not a coincidence that various forms of magic (e.g., the magical feats of Santa Claus or an evil fairy, astrology, telepathy and magical healing) have such a powerful attraction to both children[27] and adults[28,29]. The difference between a supernatural event (e.g., the ability to move inanimate objects by thought alone) and a novel natural event (e.g., the ability to move objects by means of unknown physical fields) is that *the former is not only new but also more valuable to the observer than the latter.*

In order to find out which of the two events – a magical event or an equally novel and attractive scientific event – is more attractive to children and adults to explore, an experiment was designed[30]. Participants were shown an apparently supernatural event: an object (e.g., a postage stamp), which had been placed in an empty wooden box in the participant's full view, suddenly changed its shape (e.g., became cut into parts or half-burned). One group of participants observed this event after the experimenter said a 'magical incantation', and the other after the experimenter switched on and off again an unknown physical device that produced light and sound effects. Next, the participants were asked to view this event one more time. This time, however, the object placed into the box was something of value to the participants – a nice picture that the child had been rewarded with earlier in the experiment or, in the case of adults, their driving licenses. The aim of this experimental manipulation was to find out which participants of the two groups would be more willing to run the risk of losing their valuable objects – those who were led to think they had been witnesses of a real magical event or those who believed that the event had been caused by an unknown physical device.

The results showed that a significantly larger number of both children and adults exhibited their readiness to put their valuable objects at risk in the 'magical event' group than in the 'novel physical event' group. This suggests that the tendency to explore the supernatural (e.g., reading books or watching movies with magical content) can indeed be a stronger incentive for learning than the tendency to explore novel scientific effects. Interestingly, some teachers at preschool and elementary school levels intuitively come to this conclusion by converting their learning content into the form of games with magical content. For example, a typical exercise on teaching language at primary schools in England is sentence completion: a teacher covers part of a sentence with a piece of paper and asks children to imagine that they are wizards that can see through non-transparent screens. Clearly, this way of instructing children makes them more motivated to learn than direct instruction. But is motivation to learn the only natural benefit of children's engagement with magic? Could it not be that children's fascination with magic, along with providing children with a strong motive for learning, also helps to improve their useful skills in the domains of thinking, perception and memory? In order to answer this question, we need to design an experiment.

The natural and the supernatural as stimulators of learning

Altogether, the problem of the experiment can be put as follows: observation of what kind of events – supernatural or natural ones – will have a stronger effect on children's ability to learn useful cognitive skills? For example, seeing which of the following two phenomena will help the children better reflect upon and understand the law of gravitation – an apple jumping up from the floor on its own or an apple falling down onto the floor?

At first glance, seeing the natural event might be more effective for understanding the law of gravitation than seeing the supernatural event, because the natural event conforms to the law of gravitation and the supernatural event doesn't. However, the opposite expectation is also plausible, and it is based on the fact that human thinking is not an isolated psychological faculty and works in conjunction with memory, imagination and emotions. For example, if a child is shown a natural object (e.g., an image of a horse or a bird), the child's perception and memory will be activated, but the child's thinking, emotions and imagination will remain unchallenged. The child knows what the horse and the bird are, and his or her perception simply places the familiar images onto the appropriate 'shelves' in the child's 'mental library'. If, however, the child is presented with a winged horse flying in the sky, the child's reaction is likely to be different. The sight of the supernatural creature is a challenge to the child, and the child's thinking, emotions and imagination are immediately engaged. Imagination draws contrasting natural images from the 'mental library' (i.e., the images of a horse and a bird); emotions make the child feel surprised; and thinking says: "The winged horse is a supernatural entity because horses don't have wings and are too heavy to fly." In other words, *in contrast to natural images that activate only the child's perception and memory, supernatural images activate all of the child's cognitive processes* (see Figure 8.1). Similarly, one might expect that presenting the child with the image of an apple flying in mid-air on its own would make the child

FIGURE 8.1   Imagining possible and impossible objects

reflect upon the law of gravitation and understand this law to a greater extent than demonstrating the image of an apple falling from the tree branch onto the ground.

As long as both of the above hypotheses seem plausible, only experimentation can answer the following question: which kind of phenomena – natural or super-natural ones – would have a stronger impact on children's useful cognitive skills?

Can elephants fly? Magical thinking as a stimulator of visual analysis

To compare the impact of natural and supernatural entities on children's cognitive abilities, it is necessary first to establish whether children can understand the diffe-rence between the natural and the supernatural. Research has shown that three- to five-year-old children poorly distinguish between pictures of real and fantastical animals[31]; only 11-year-olds proved able to contrast real and fantastical animals com-parably to adults[32]. The question arises as to why preschool and elementary school children find it so difficult to distinguish pictures of real animals from those of fan-tastical ones: because they cannot compare the pictures between themselves (deficit of visual analysis skills) or because they don't understand the difference between fantastical and real animals (deficit of understanding the difference between fantasy and reality). In order to answer this question, researchers replaced the pictures with verbal descriptions of real and fantastical entities. For example, children were asked whether a certain entity (e.g., a clown or Santa) could be in many different places simultaneously. It turned out that when visual analysis was unnecessary, even four- and five-year-olds were able to successfully distinguish between real and fantastical entities[33]. This study suggested that children at the age of four years begin to under-stand that fantastical entities have abilities that violate the known laws of physics, biology and psychology. The fact that children below the age of 11 years struggle to distinguish between pictures of real and fantastical entities can be explained not by a lack of understanding of the difference between fantasy and reality, but by an immaturity of visual analysis skills.

One way of improving children's ability to visually differentiate between pictures that show natural and supernatural entities could be exposing chil-dren to a movie with magical entities and events. As argued above, watching a supernatural entity (e.g., a flying elephant) makes children visually represent the contrasting natural entities (e.g., an ordinary elephant and a regular bird) and reflect upon the visual features that distinguish the supernatural (fantastical) entity from its natural (real) counterpart(s). In contrast, observing a natural entity (e.g., an ordinary elephant) doesn't provoke the child to make such a comparison; indeed, it only makes the child recognise the given animal (e.g., an elephant) and keep it for a time in his or her working memory. In other words, watching a movie with supernatural entities is an exercise in the ability to distinguish between visual images of the natural and the supernatural, whereas watching a similar movie with natural entities is simply training the ability to recognise visual images of familiar objects.

In order to find out whether watching a movie with magical content will indeed enhance children's abilities of visual analysis to a larger extent than watching a similar movie without magical content, participants (six- and nine-year-old school children from Lancaster, England) were shown two video clips. Both video clips contained extractions from the movie *Harry Potter and the Philosopher's Stone*. The supernatural clip contained phenomena such as riding a broomstick, an intelligent serpent, people becoming invisible and other events that violate the known principles of physics, biology and psychology. The natural clip contained the same characters as the supernatural clip, but no supernatural magical events. Eight independent raters scored the clips on such scales as interest, visual and sound effects, speed of events changing and presence of supernatural events. On all of the scales except presence of supernatural events, both clips were given approximately equal scores. Children in the experimental group were shown the supernatural clip and children in the control group were shown the natural clip.

Before and after exposure to the movie clips, the children were administered an interactive computer test of their ability to distinguish between supernatural and natural characters and events[34]. In the test, a child was shown a display that contained a picture at the top and two pictures at the bottom. The pictures portrayed natural and supernatural characters and events. The child was instructed as follows: "Look at the picture at the top of the screen and decide whether such an event (a creature) can or cannot be in the real world. Out of the two pictures at the bottom of the screen click on the one that matches the picture on the top." There were 42 trials altogether. The test assessed the child's ability to distinguish natural (not fantastical) entities from supernatural (fantastical) ones.

The results indicated that before exposure to the video clips, the children of both groups scored approximately equally in terms of their ability to distinguish between natural and supernatural visual displays; after the exposure, children in the experimental group scored significantly better than before the exposure, whereas children in the control group did not[35]. The study supports the hypothesis that *exposure to supernatural magical creatures and events facilitates children's ability to distinguish between fantastical and realistic visual displays, whereas exposure to natural creatures and events does not*. This confirms that engaging children in magical thinking can be used as a method of improving children's visual analysis skills. This ability is necessary for most modern educational methods that involve computer graphics and other means of visual representation of content.

Magical thinking as a stimulator of creativity

In another study, the task was to examine the impact of watching a movie clip with supernatural entities and events on children's creative thinking by comparing this impact with the impact of watching a similar movie clip that contained no supernatural entities and events. *Divergent creativity* is the ability to find new, non-standard ways of solving a problem that allows for multiple solutions. Since supernatural events violate the known laws of science and thus create new and original ways of solving familiar tasks, *observing supernatural events may be a way of facilitating divergent creativity*. For example, if the task is to move from one city to another, then the ordinary ways of solving this task could be going by train, by car or on a horse; but when children are shown a movie in which people use a dragon, a magical carpet or a broomstick in order to move from one place to another, this can make the children start searching for unusual ways of problem solving and thus stimulating their divergent creativity.

Accordingly, the aim of the study was to find out whether exposing children to a movie clip with supernatural events will affect their creative thinking to the same or greater extent than exposing them to a movie clip with equally interesting and engaging natural events. Four- and six-year-old children from a primary school in Greater London were divided into experimental and control groups. The experimental group was shown a movie clip with magical events and the control group was shown a movie clip without magical events; the movie clips were matched for interest and visual and sound effects. Before and after exposure to the movie clips, the children were individually tested on divergent creativity. One of the tests was Torrance's 'Creativity in action and movement test', which included tasks such as inventing as many natural uses for a paper cup as possible (e.g., as storage for pencils)[36]; the other test was drawing a supernatural entity (a plant, a car or a person)[37]. The results indicated that before exposure to the films, the children in both groups exhibited creativity to an approximately equal extent; after exposure, children in the experimental group scored significantly higher on most creativity measures than children in the control group (see Figure 8.2). In another version of this experiment, the same method was used, with a few minor modifications. Along with being tested on creativity, six- and eight-year-old children from a primary school in Shropshire, England, were also tested on their belief in the reality of magical events. Like in the previous experiment, after but not before exposure to the movie clips, children in the experimental group exhibited greater creativity than children in the control group. However, there was no significant difference between the groups on children's belief in the reality of magical events[38].

Altogether, both experiments showed that *exposing children to a movie clip with magical events increases creativity in children to a significantly larger extent than exposing the children to a movie clip with equally exciting but non-magical events*. At the same time, the experiments indicated that exposing children to a movie clip with magical events did not increase children's belief in the reality of magical events. This suggests that exposure to supernatural events can be used for educational purposes without such exposure increasing children's belief in magic.

Magical thinking as a stimulator of memory

One more study compared the effects of the natural and the supernatural on memory[39]. The study was based on the fact that TV adverts often include scenes that violate the known laws of physics, biology and psychology (e.g., a piece of chocolate turns into a little person or a moving car becomes a running jaguar). Arguably,

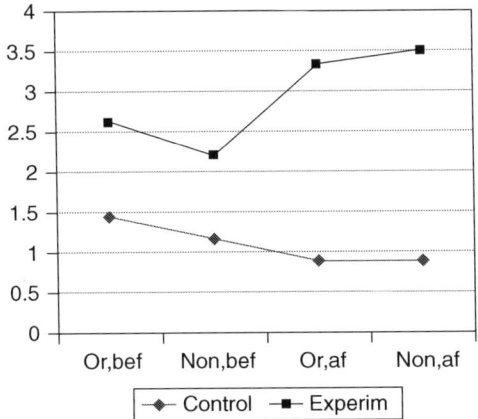

Proportions of scores for drawings' originality
and non-existence, as a function
of Condition (experimental vs. control) and Time
(before vs. after intervention)

FIGURE 8.2   The effect of watching video clips with (upper line) or without
(lower line) magical content on creativity of children's drawings. Or = originality;
Non = non-existence; bef = before exposure to the movie clip; af = after exposure to
the movie clip
Source: Subbotsky, Hysted & Jones, 2010

this is done under the assumption that advertised brands will be remembered better
if they are placed in the context of magical creatures and events than if they are
placed in the context of non-magical creatures and events. However, thus far, there
has been no experimental verification of this assumption.

Accordingly, the aim of the study was to find out whether commercial products
placed in the context of supernatural events are remembered better than similar
products placed in the context of natural events. British adolescents and adults were
shown two films composed of TV advertising clips. Film 1 included clips advertising commercial brands (e.g., Mini Cooper, Levi's, Paco Rabanne and Pepsi Max)
put in the context of supernatural events, and Film 2 had similar brands placed in
the context of equally interesting and exciting natural events. Independent scorers
assessed the films on such scales as presence of magical effects, interest, emotional
attractiveness and visual and sound effects. The films did not differ significantly one
from the other on any of the scales but presence of magical effects. Participants were
individually shown both films (the order of the demonstration was randomised)
and then asked to recognise advertised brands from a list of brands in which the
advertised brands were randomly mixed with similar brands that had not been
included in the films. For every correctly recognised brand a point was awarded and
for every incorrectly recognised brand a point was deducted.

The results indicated that both adolescents and adults recognised a significantly
larger number of brands from the film with a magical context than from the film

with a non-magical context. This confirmed the assumption that *associating advertised commercial brands with magical events helps to increase subsequent recognition of these brands by the viewers*. A plausible explanation of this effect is that supernatural events engaged the participants' emotions, thinking and imagination and thus helped to imprint the brands associated with these events in the participants' memory. In contrast, natural events did not stimulate the participants thinking and imagination to the same extent as the supernatural events; as a result, the participants' recognition of commercial brands associated with the natural events was not as good as their recognition of the brands associated with the supernatural magical events.

Conclusion: what can magic teach us?

This chapter began with the following questions: do children really believe in magic or do they only pretend that magic exists in the real world? Is children's interest in magical phenomena stronger than their interest in new physical phenomena? Could children's magical beliefs be an impediment to their science education? Could magical thinking be used to improve children's cognitive skills?

The reviewed studies suggest that preschool and elementary school children do indeed believe in magic. These children also show a considerably stronger interest in exploring magical events than in exploring new physical events. Further, it turned out that exposure to a movie clip with supernatural events stimulated some of the children's cognitive skills to a significantly greater extent than exposure to a movie clip with equally interesting and exciting natural events. Finally, commercial brands placed in the context of supernatural magical events were subsequently recognised significantly better than commercial brands placed in the context of natural events. The explanation offered for these effects was that *exposure to the natural engages a narrow circle of psychological functions (mostly perception and working memory), whereas exposure to the supernatural activates a whole spectrum of psychological functions (perception, memory, emotions, imagination and thinking) and thus facilitates the children's cognitive skills to a significantly greater extent than exposure to the natural.*

English literature for children often involves magical thinking. For example, Raymond Briggs' book *Fungus the Bogeyman*[40], recommended for reading in primary school Year 2, portrays a magical character 'Bogeyman' (a humanoid mould), which has an unusual body structure, sleeps only in a dirty bed, keeps its clothes and shoes in cold water and generally does everything the opposite to how things should be done. The studies reviewed in this chapter suggest that such intuitive uses of magical thinking by writers are indeed psychologically justified. In principle, *it is possible to create alternative handbooks on scientific subjects in which familiar things would follow not the laws of nature, but the laws of magic instead.* These alternative handbooks would not replace regular handbooks, but rather complement them, helping students better reflect upon the laws of nature via the 'compare and contrast' method[41,42]. Jean Effel's *The Creation of the World* cartoon book could be an example[43]. The biblical story of how the world was created is different from the story science tells us, yet these two stories share some common features. For example, the Big Bang theory is similar to the biblical story in that it portrays the universe as appearing from virtually nothing (the initial singularity) and expanding very fast; this rapid expansion conforms not to the laws of physics, but to the laws of magic[44]. Discussing with children common features and differences between scientific and biblical versions of the origins of the universe or the origins of species might make science lessons more interesting.

In reality, such alternative 'handbooks' already exist, though not in real life, but rather in fiction books and movies. For example, some parts of the *Harry Potter* series can be used for teaching science. When watching or reading through the scene in which Harry and his friends go through a stone wall in order to reach the world of magic, children could be asked how the stone wall could let people through while not letting light through at the same time. When watching the scene in which Harry wears the cloak of invisibility, it might be interesting to discuss whether the hero himself could see anything at all; indeed, in order to be invisible, a person has to be transparent and let all light through his or her body, but if the person's eyes did not absorb light, the person wouldn't be able to extract any information from the light and therefore would be unable to see anything. When the characters ride broomsticks, it could be interesting to ask children how it is possible to maintain equilibrium at turning points while sitting on a narrow rod and without having any fulcrum; it might also be useful to discuss issues such as gravitation, inertia and centrifugal forces. When Harry speaks Parseltongue to a boa constrictor, one could discuss whether the communication is happening through sound, visual signals or telepathically via a direct 'mind-to-mind' transfer of messages. Discussions like these could help children reflect upon the known laws of physics, biology and psychology better than just direct instructions in those laws.

So, does children's early belief in magic inhibit their ability to learn the laws of science? In light of the studies reviewed in this chapter, the answer is no. The belief in magic and the belief in science are not mutually exclusive; they can peacefully coexist in one mind. Consciously, most modern adults believe in science and not in magic. However, in specially designed psychological experiments[45] or in some existential situations of high stress[46], a rational person can discover that his or her early beliefs in magic and miracles are still there. Research has shown that adults' subconscious belief in magic coexists with their conscious belief in science and performs certain important functions[47]. As for children, their early belief in magic and fascination with magical things could be used to enhance their cognitive development. Far from being just a 'wrong vision of reality', children's magical thinking opens up new and exciting opportunities for improving cognitive skills and making science education more effective and interesting.

Notes

1 Bettelheim, B. (1977)
2 Bloom, P. & Weisberg, D. S. (2007)
3 Hsu, C. T., Jacobs, A. M., Altmann, U. & Conrad, M. (2015)

4 Dias, M. G. & Harris, P. L. (1988)
5 Hawkins, J., Pea, R. D., Glick, J. & Scribner, S. (1984)
6 Leevers, H. J. & Harris, P. L. (1999)
7 Lillard, A. S. (1996)
8 Lillard, A. S. & Sobel, D. (1999)
9 Principe, G. F. & Smith, E. (2008)
10 Richards, C. A. & Sanderson, J. A. (1999)
11 Sobel, D. M. & Lillard, A. S. (2001)
12 Richert, R. A., Shawber, A. B., Hoffman, R. I. & Taylor, M. (2009)
13 Richert, R.A. & Smith, E. I. (2011)
14 Atkin, C. (1983)
15 Bushman, B. J. & Huesmann, L. R. (2001)
16 Piaget, J. (1929/1971)
17 Subbotsky, E. V. (1984)
18 Harris, P., Brown, E., Marriot, C., Whittal, S. & Harmer, S. (1991)
19 Rosengren, K. S. & Hickling, A. (2000)
20 Woolley, J. D. (2000)
21 Woolley, J. D., Boerger, E. A. & Markman, A. B. (2004)
22 Lepper, M. R., Corpus, J. H. & Iyengar, S. S. (2005)
23 Mendel, G. (1965)
24 Henderson, B. & Moore, S. G. (1980)
25 Nemeroff, C. & Rozin, P. (1992)
26 Pliner, P., Pelchat, M. & Grabski, M. (1993)
27 Johnson, C. & Harris, P. L. (1994)
28 Jahoda, G. (1969)
29 Luhrman, T. M. (1989)
30 Subbotsky, E. (2010a)
31 Taylor, B. & Howell, R. (1973)
32 Morrison, P. & Gardner, H. (1978)
33 Sharon, T. & Woolley, J. D. (2004)
34 Subbotsky, E. (2012)
35 Subbotsky, E. & Slater, E. (2011)
36 Torrance, E. P. (1981)
37 Karmiloff-Smith, A. (1989)
38 Subbotsky, E., Hysted, C. & Jones, N. (2010)
39 Subbotsky, E. & Matthews, J. (2011)
40 Briggs, R. (2012)
41 Marzano, R. J. (2007)
42 Marzano, R. J., Marzano, J. S. & Pickering, D. J. (2003)
43 www.ebay.fr/itm/Jean-Effel-The-Creation-of-the-World-book-cartoon-caricature-226-pg-1963/272365953720
44 http://en.wikipedia.org/wiki/Big_Bang
45 Subbotsky, E. V. (2001)
46 Keinan, G. (1994)
47 Subbotsky, E. (2014)

9
GAMES WITH THE SUPERNATURAL

Magical reality in everyday life

Problem

One of the functions of children's role play is overcoming the inferiority complex: in play, children temporarily forget that they are small and weak and get the opportunity to rebuild the world as they please by affecting people and objects in a magical way. The computer games Minecraft and Roblox are examples. By immersing themselves in these games, children obtain powers that are truly magical: with a motion of a finger they can create cars, spaceships, gardens, palaces and whole cities, both on land and underground. In endless labyrinths of underground tunnels the children can travel, chase and kill bad guys or hide themselves from monsters. All is accompanied by nice music; all is full of bright colours and shades of grey. Pulling the children out of this magical world and back into the real world, where every achievement requires efforts and they must obey rules, brings disappointment to the children. Quite understandably, they protest.

As a result, there appeared a new psychological phenomenon: addiction to the magical reality of computer games. In some respects, this addiction is similar to the addiction to psychedelic drugs. In the altered state of mind caused by LSD, a person can get the intensive experience of the freshness of the world and feel a oneness with surrounding objects. While playing computer games, children may have similar experiences. It is quite possible that in this state of consciousness the children learn faster and remember new ideas better than in the normal state of mind. But addiction to the digital magical reality is also potentially dangerous. This addiction affects the children's thinking style: children become habituated to receiving information in the form of chunks of rapidly changing visual images instead of distinctive symbolic structures such as written words and mathematical equations. The number of children who enjoy reading books is plummeting. Besides, we don't know yet whether this addiction might cause some delicate changes to the brain chemistry and what the long-term consequences of this addiction might be. It is possible that the World Health Organization will recognise 'gaming disorder', which includes addiction to computer games, as a mental health condition in its next revision of the International Classification of Diseases[1]. This may or may not happen, but one thing is clear: in children's lives, *there has appeared a new and unexplored activity that can plunge children into the depths of magical reality faster and more effectively than traditional ways of playing with magic – fairy tales and pretend games.*

Computer games are just one type of play with magical reality. Intentionally or unintentionally, in modern industrial cultures people play a variety of games with magical reality on a regular basis. The easy access of modern people to magical reality, which is no longer controlled by religion, raises important questions. On what domains of modern life does magical reality trespass? What motivates people who live in the world designed by science to play with magical reality? What psychological consequences could the involvement with magical reality entail? These and other issues will be discussed in this chapter.

Wonderland

Magical reality, which in this chapter will be referred to as Wonderland, is similar to the geometry of a sphere. In such a geometry, the postulates of Euclidean geometry are suspended: parallel lines can cross and the sum of the angles of a triangle can be more than 180 degrees[2]. Likewise, in Wonderland, the known laws of logic and nature can be violated: a part of a whole can be equal to the whole, a statement can be both true and false, time can go backwards, horses can have wings and fly, people can travel to other universes and little children can solve advanced problems of high maths. But Wonderland is not a land of chaos: it has laws and phenomena of its own. One of the laws of Wonderland is 'the law of contagion'. According to this law, two objects that used to be in contact but at the moment are physically unrelated to one another can have a magical bond between them that lasts forever and acts instantly at a distance. For example, if a wizard casts a magic spell and burns a piece of clothing of the person to whom the wizards intends to inflict harm, then the target person may become ill or even die. In Wonderland, a thought can instantly come true (a direct 'Self over matter' magical phenomenon). In everyday reality, if we decided to build a house, we have to get some bricks and other construction materials, make a plan, lay down a foundation and do other things before the house can take its physical form. By contrast, in the supernatural reality of Wonderland, we can build a palace by saying a magic spell and wishing the palace to appear (see Table 4.1 for more on the laws and phenomena of magic).

Sometimes we get into Wonderland in our dreams. Indeed, in dreams, we can see our deceased relatives, speak with animals and fly in the air like birds. But many of us like to play with the supernatural in the waking state of mind as well. We enjoy watching films with magical content (e.g., *The Lord of the Rings* or the *Harry Potter* series), reading books that depict magical events (e.g., *The Master and Margarita* by Mikhail Bulgakov) and looking at art objects that include magical characters (e.g., paintings by Dali, Picasso or Magritte). The name for this type of playing with the supernatural is *magical thinking*. Because magical thinking unfolds within the domain of the imagination, it peacefully coexists with our belief in science. In the Western world today, most people view the belief in Wonderland to be a remnant of ancient history; in the popular view, only small children and a limited number of superstitious adults take magic seriously.

However, psychological studies of recent decades have shown that, deep in the subconscious, educated adults still believe in the supernatural[3,4]. For example, according to the law of contagion, a warrior's weapon is magically linked to its owner and can pass the owner's power and military skills to another person who takes possession of the weapon. An example can be found in Russian folklore, where a sword that belonged to a great warrior kept the warrior's power even after the warrior's death and could pass this power to a new owner. With the aim of examining whether this law works in the minds of modern educated adults, in a recent study, participants (university undergraduates) were given a golf putter and told that a professional golfer had owned it[5]. The results indicated that these participants not only were more successful at putting the ball into the golf hole than the participants who had been told that their putter had just been purchased in a store, but they also perceived the size of the golf hole to be larger. This shows that the belief that a tool can magically transfer the skill of its owner to another person is indeed present in the minds of modern adults; more important, this belief works by facilitating the performance of those who think that their tool previously belonged to a person who had used this tool successfully. Other studies, reviewed in the previous chapters, showed that in certain conditions modern educated adults who initially denied their belief in magic changed their minds and consciously acknowledged their belief in the supernatural over the course of the experiments[6,7].

Interestingly, people can attribute magical powers not only to professional witches, but also to other influential individuals, such as scientists, politicians and medical doctors. This subconscious belief of modern people in magic opens up opportunities for manipulating people's minds with the aim of extracting political, social and economic gains (see Chapter 7). Let us consider how manipulating the mass consciousness based on people's magical beliefs works in some domains of modern life.

The magic of economics

The popular sentiment 'free cheese exists only in a mousetrap' is not a joke but a law of economics, as unbreakable as the law of gravity in physics. Whatever nooks and crannies of Wonderland modern virtual realities bring us to, one thing stays unchanged – the 'exchange principle'. One can ride a broomstick, be invisible, travel back and forth in time or turn people into stone, but one cannot get anything from another person without giving the person something in return, even if this something is the donor's feeling of moral satisfaction. Still, just like in dreams where we sometimes cheat the law of gravitation, it seems to us that we can cheat the law of exchange and get the free cheese while avoiding the mousetrap.

Every year, people are presented with dozens of offers from companies: buy our goods and your name will be put in a lottery where you could win thousands of pounds. I used to bin such offers, but once I decided to give it a try – just to see what would come of it. Coincidentally, the choice of goods was not too bad. I purchased three small but useful objects and waited for my lottery win. Of course, I won nothing, but the offers from the same company kept coming, accompanied by the same catalogue. The only small variation happened when the offer came for the fourth time; along with the catalogue of goods for sale and a standard set of papers to complete, a small poster came with my name and the company's gold medal shown on it. The instruction said, "Proudly put on the wall" – a small carrot with the aim of making the suggestion work again. It reminded me the children's story, Pinocchio. The marvellous but naïve puppet boy named Pinocchio listened to the crooked cat and fox and planted his coins in the Miracle Meadow, hoping that a tree would soon grow, dripping with gold coins. Many people in 1990s Russia, including scientists with PhDs, lost their savings in treacherous financial pyramid schemes that promised unrealistically high rates of capital growth. What does this fact prove? A simple truth: *science education cannot protect a person from deception that targets the person's subconscious belief in the financial magic of the gurus of business* (see Chapter 7 for a more detailed account of the psychological causes of manipulation of the popular consciousness).

Recall that according to the law of sympathy, things that resemble each other are magically linked to one another (see Table 4.1). For example, a rhinoceros horn looks like a penis and is used in some cultures as an aphrodisiac. In 2002, American psychologists Daniel Kahneman and Amos Tversky studied cognitive biases in economics[8]. One such bias is the anchoring effect. This effect describes the common human tendency, when solving economic tasks, to rely on information that becomes a frame of reference despite it having no rational connection with the task's essence. For example, if a person is asked to select a two-digit number from a pool of numbers and then estimate the approximate cost of various goods (e.g., a brand of wine, a chocolate bar or a computer), then the person who chooses a larger-digit number would estimate the cost to be higher than the person who chooses a smaller-digit number[9]. Why do people do this? *Because unconsciously they follow the law of sympathy, according to which things tend to generate other things that are similar to them.* If a person selected the number 11, then in the person's mind, a bottle of wine simply can't cost £50, but if the person selected the number 60, then it can. That is why it's easier for manufacturers to sell a dress that costs £160 if the dress is labelled as having gone down in price from £200 to £160 than if the label only shows the real price of £160.

The aforementioned 'law of contagion', which works mostly subconsciously[10], can also be used in order to handle a harsh situation in the economy. In his book about the financial crash of 1929, American economist John Kenneth Galbraith noted that during the crisis, President Hoover deliberately employed the law of contagion to elevate the spirit of the population by conducting meetings that had no practical significance[11]. He invited to the White House various VIPs (e.g., governors of the states, manufacturers, trade union leaders, businessmen, politicians, etc.) who discussed issues of no significant value. Nevertheless, *in the eyes of the general public, the impression built up that some important activity was going on, because the importance of the persons involved in these meetings was magically transferred to the meetings themselves.*

In the same book, Galbraith maintains that in moments of crises and those closely preceding crises, the influence of 'economics gurus' – the authorities in business – on the rise or fall of share prices rapidly grows. For instance, when director of General Motors John Raskob, while departing from USA to Europe in 1928, predicted the rise of General Motors' share price in the next few days, the share prices of this and other companies indeed went up sharply the next day. In March 1929, a single optimistic prediction of the director of National City Bank John Mitchell helped to inhibit the fall of share prices. *What affects the market are not rational calculations, but public belief in the special abilities of prominent financiers.* The science-based predictions of Harvard University professors of economics had a smaller effect on the public than the arbitrary declarations of big figures in business. Interestingly, as the crisis deepened in the following years, nobody accused the bosses of business of wrongly optimistic predictions that resulted in hundreds of thousands of people losing their fortunes; in contrast, the Harvard professors' reputations were irreparably damaged.

Galbraith himself suffered from the financial Wonderland. He published his book on the 1929 financial crash in 1955. When somewhat later that year the financial market collapsed again, some members of the public accused Galbraith's book of this. Galbraith started to get letters with threats of physical violence and promises of prayers for his early death and deterioration of his health. *About that time, Galbraith had a skiing accident and broke his leg; some of his accusers took this event as proof of the effectiveness of their prayers.* The belief in the direct 'Self over matter' magical phenomenon of Wonderland played out with the perseverance of a machine.

The laws of Wonderland are particularly influential in the area of advertising. Studies have shown that placement of commercial products within films positively affects viewers' ability to remember these products; if a movie character personally uses a product (e.g., pouring Evian water into a plate in order to give it to a dog), then this increases viewers' general positive attitude towards that product[12]. But why do people develop a more positive attitude towards a product used by a movie character than towards other similar products placed in the context of the same movie? Because the movie star touches the product, and people want to be as beautiful and famous as the movie star. The magical laws of contagion and sympathy are at work here: *an object that was in physical contact with a celebrity magically absorbs the qualities of the celebrity, becomes 'contaminated' with these qualities and then passes these qualities on to those who possess a similar object.* Subconsciously, viewers hope that if they use the same product (e.g., Evian water), then some of these qualities will pass to them.

The magic of numbers dates back to Pythagoras (570–490 bce), but it also works today. Among magical numbers, the number three is perhaps the most popular: remember the three little pigs of the popular children's tale, the three musketeers and the three magi of the nativity. A recent psychological study has shown that advertising is affected by the magical influence of the number three. If you want a product you advertise to be viewed favourably by potential consumers, you need to name three positive qualities of the product, and it is irrelevant whether the advertised product is a flask of shampoo, a hotel or a presidential candidate[13]. But if you try too hard and name four or more positive qualities, the magic disappears and the advertised product loses its charm in the eyes of consumers. Do you want to positively present a man? 'Handsome, intelligent and nice' would do the job, but 'handsome, intelligent, nice and a good sportsman' would be an overshoot. But why do people trust only three adjectives? Could it not be because the number three comes from Wonderland and is a magical number? Beside its unique mathematical properties, the number three is fundamental to many world religions, such as Christianity, Buddhism, Hinduism, Taoism and Wicca. In Chinese and Vietnamese mythology, the number three is a lucky number[14]. *Guided by the magical law of sympathy, people transfer their trust in the number three to products that have three positive adjectives.*

The commercial advertising that interrupts a programme every 15 minutes annoys many people who watch TV. But if you look at advertisement clips more closely, you might notice that some of them involve a good knowledge of human psychology. I was always interested in watching advertisements with magical effects. Why do advertising companies need such effects, which are cognitively complex and probably expensive to make? Do they assume that such magic-clad products are remembered better and liked more by potential consumers than products placed in a non-magical context? Research reviewed in the previous chapter supported this assumption. It turned out that brands placed within a magical context indeed were better recognised by viewers in a subsequent test than brands placed within a non-magical context[15]. *By the law of contagion, commercial brands 'stick' to the attention-grabbing magical effects and are therefore remembered better than brands clad in an attractive but more mundane visual context.*

The magic of medicine

Modern scientific medicine has been around for approximately 250 years, but how did people live without it in earlier times? They believed in the healing powers of medicine men, shamans and priests, and this belief was sufficient for a healing effect to take place. This magical effect of a patient's belief on his or her physical state is known as the placebo effect, and this effect is still used in medicine today. To achieve the placebo effect, a doctor gives a patient a capsule filled with distilled or slightly sweetened water and tells them that this is the medicine for the patient's illness. It turns out that the patient's belief that the doctor is telling the truth can be enough to produce some improvement in the patient's condition[16]. Recent studies have shown that even when patients are told that the capsule they are given contains distilled water but that taking the capsule could nevertheless help, a healing effect can still follow[17]. In other words, the belief of the patient in the doctor's healing powers makes the patient's brain elicit substances (hormones and opiates) that can, albeit temporarily, reduce the intensity of pain and other symptoms. Of course, the placebo effect cannot cure serious illnesses, such as cancer or tuberculosis, but the fact remains that our thoughts alone can influence physiological processes in our bodies.

According to some data, up to 35% of medical doctors today occasionally use the placebo effect in their clinical work[18]. The advocates of magical healing methods widely employ the placebo effect as well, and it isn't surprising that in some cases magic really does work. However, unlike medical doctors, magical healers are not bound by the Hippocratic Oath of 'do no harm' and may use the placebo effect in circumstances where this effect is insufficient for cure (e.g., in cases such as cancer, Alzheimer's disease and Parkinson's disease). This provokes unrealistic hopes in the patient and, as a result, the patient may apply for proper medical treatment when it is too late.

Homeopathy is another baby of sympathetic magic. In accord with the magical law of sympathy, homeopathic medicine relies on the principle that 'similar should be treated with similar'. The idea is that the 'medicine' should contain a substance that causes symptoms similar to those caused by the illness, but in a very diluted form. This medicine is dissolved in water or spirit, then some water is added and the solution is stirred, then the procedure of adding water and stirring is repeated many times. Homeopathic doctors believe that the resulting solution works better than those recommended by official medicine. The mechanism of the healing effect is that the water is 'charged' by the energy of the original substance, plus the energy of the homeopathic doctors themselves. According to some data, in European countries, up to 40% of practicing doctors employ some homeopathic methods. Clearly, official medicine and chemistry cannot account for the mechanisms of 'charging' water with energy, but even if the healing results of homeopathic medicine do not exceed the results of the placebo effect, homeopathic healing might still have some healing effect. No wonder that in 1999 alone in the USA, a few million people attended homeopathic doctors[19].

In order to avoid lethal mistakes, in the twentieth century, magical medicine started to present itself not as an alternative to scientific medicine, but as medicine that complements scientific medicine. Few people would argue against the idea that when a patient has terminal cancer and all of the possible scientific methods of treatment have already been used, it is necessary to raise the patient's spirit; then the rituals of 'protective magic' might help[20]. This is a no-lose situation: if the magical rituals don't help, then they don't do any harm either, but what if they do help?

It turns out that magical medicine could even be used to 'cure destiny'. Everyone knows that destiny can't be changed. Yet recently, Russian TV showed a programme about a form of surgery in Japan in which a person's destiny is corrected by carving new paths on the person's palm. And the number of people who would like to 'correct fate' grows. Obviously, these people believe that the lines on a person's palm reflect the person's fate laid down in Wonderland. Yet, speaking rationally, to believe that by changing these lines one can change one's fate is the same as believing that one can change a riverbed by altering the river's track as shown on a map. But rational logic doesn't work in Wonderland. In Wonderland, causality works both ways: for those who believe that palm lines and fate are magically interconnected, this single fact is sufficient for the hope that changing one of the pair might change the other. An interesting question for the believers might be: if an angel (or a demon) who lives in Wonderland changed our fate, would our palm lines change as well?

The magic of politics

Imagine that in the well-known biblical story God did not part the waves of the Red Sea in order to let the Hebrew people cross the sea and avoid being caught by the pharaoh's army; instead, Moses taught the people to build large rafts and the people used the rafts to cross the sea. Imagine also that Jesus did not raise Lazarus from the dead; instead, he cured a badly ill yet still living Lazarus. Would reducing miracles to a lower rank of rare yet possible events affect the image of political and religious leaders and diminish these leaders' unconditional respect and authority among their peoples? It is quite likely that it would.

The question is therefore why gods need to be mighty wizards and not just magnified copies of ordinary people. Who need gods to be wizards – gods or people? It looks like people need this more than gods. First, they need this in order to explain the unexplainable – the origins of the world and the human soul. But most important, people's leaders need this in order to sanction their political and spiritual power. Indeed, in the animal kingdom, a leader (e.g., an alpha male) enjoys its authority only temporarily and has to permanently fight for it. In contrast, in a human society, a leader can enjoy his or her authority for life and pass their title to his or her heirs. But for this to be possible, the leaders have to make the people believe that they (the leaders) are under the protection of the gods. For instance, early Egyptians believed that the pharaoh was the son of a god. The authority of modern spiritual leaders (e.g., the Pope) is sanctioned by God as well. Finally, as will be argued later in this chapter, the advantage of good over evil is impossible to maintain without the approval of the gods.

The assumption that modern people believe in the supernatural powers of their leaders explains the effectiveness with which leaders can manipulate the mass consciousness. The people of Japan believed that their emperor Hirohito (1901–1989), who ruled in the time of World War II, was a descendant of the gods[21]. This popular belief contributed to the fanatical stubbornness of the Japanese army's resistance during its battles with the US army. In the Battle of Okinawa (April 1945) Japanese suicidal pilots (kamikazes) inflicted heavy casualties on the American fleet, and it became clear that breaking the Japanese army's resistance with regular weapons could only be done at great cost. The awareness of this was a main factor in American's decision to use the atomic bomb. A kamikaze ('divine wind' in Japanese) was a volunteer who sacrificed his life for the divine values and expected rewards in the afterlife[22]. Studies suggest that the belief in divine values is the foundation of modern political terror as well. For example, Palestinian suicidal terrorists didn't differ from their compatriots in any way except for their fanatical belief in God[23]. *The belief in the supernatural helps a suicidal terrorist to overcome the fear of death, which in ordinary people serves as a block that prevents the release of the energy necessary for committing an act of self-destructive terror.*

In Chapter 7, we discussed how the ideologists of the free market economy in Russia managed to persuade the workers that the privatisation of plants and factories was in their own interests; this was done by sheer manipulation of the workers' consciousness and without giving any proof of any benefits whatso-ever[24]. This manipulation was, however, carefully prepared. In the middle of 1980s, Soviet central television gave its main channels over to psychic healers – Anatoly Kashpirovsky and Allan Chumak – who practiced seances of magical healing with millions of people watching. When the TV screens suddenly swapped magical healers for the ideologists of privatisation, the people's consciousness was already 'softened' and the psychological ground for the suggestion ready. By taking advantage of this moment, Boris Yeltsin's economists easily 'persuaded' the people that privatisation would magically improve their economic conditions.

In a similar vein, many Western nations today unquestioningly follow their leaders' appeals to initiate local wars and change regimes in independent countries, even when in private many people understand the futility of such enterprises and are sorry for the loss of lives of their soldiers and the bloodbaths that usually follow the toppling of unwanted regimes. *This apparently illogical behaviour of educated people should not be surprising if we accept that people's subconscious beliefs in the godlike rightness of their political leaders usually override rational considerations.*

The magic of morality

"I want to be a mean boy," five-year-old boy once told me, having watched an American cartoon with a teenager named Ben Ten as the main character.

"Why?" I asked.

"Mean boys are strong," the boy answered. And I confess, I struggled to find a persuasive response. The boy was still too young to grasp the idea of the rewards and punishments that might await a person in the afterlife. How could I prove to a five-year-old person that good is better than evil? Indeed, Ben Ten is not a kind boy, but he is strong, witty, brave and always wins. In his classroom, the boy sees the same: strong, bold and mean boys take the upper hand, whereas kind and gentle boys try to copy them. On TV we often see lions, tigers and crocodiles catching and eating beautiful little babies of wildebeests and zebras. The number of documentaries about Hitler dwarfs the number of films describing the lives of Mother Teresa or Albert Schweitzer. Trying to scare the boy by earthly sanctions that might follow his mean deeds (e.g., punishments that he might get from his parents or teachers) is not the right way either. But what is the right way? Where could I find a proper argument for the advantages of good over evil? All I could do was try to get help from the boy's magical thinking. My assurance that Santa Claus doesn't bring nice presents to bad boys seemed to make an impression, but would this effect last long? And what if the boy discovered that Santa Claus does bring presents to bad boys after all?

Indeed, what is good? What is evil? If we view other people as beings equal to ourselves, then we must follow the Golden Rule of morality: 'treat others as you want to be treated'. But if we view others as beings inferior to us (as we do indeed view some other creatures; e.g., domestic animals), then we may do with those others as we please. Let us call the belief in equality between people 'good' and the belief in inequality 'evil'. Evil justifies all sorts of manipulations of other people: slavery, captivity, rape, sadism and murder. But what makes some people believe in good and the others in evil? The popular answer is that people believe in good because if it were otherwise then humankind would not have survived and people would have exterminated each other in a constant war of 'all against all'. But this is not so. In the animal kingdom, there is no good or evil, but predatory animals do not exterminate each other to extinction. A lion that fights with another lion would not kill the opponent if the opponent runs or takes a posture of submission. A human person does not have similar instincts, but humans do have another mechanism that stops them from following evil: the socially developed ban on killing tribesmen.

The real problem is to understand what goes on inside our minds when we are free from surveillance. Suppose we observe an evil action that stays unpunished or is even rewarded. We find ourselves in a situation like that when we watch films or read books such as *The Talented Mister Ripley*. In the depths of our minds, where no one can see us, we are free to condemn or approve evil. Films like that stir our feelings as we unwittingly ask ourselves, "If I were in the character's shoes, what would I do?" Indeed, our primordial instinct, which we inherited from our animal ancestors, dictates us to choose a mean but profitable action, because in nature the strongest and the fittest survive, and choosing the profitable action increases our chances of survival. And now, having watched the film, inside our minds we feel as if two ropes are pulling us in opposite directions: our innate selfishness pulls us towards evil, but what pulls us towards good? There are no innate genetic mechanisms that force a person to do morally good things, because if such mechanisms existed, everyone would be a moral person. There are of course innate empathic feelings, but these feelings are weak and depend on circumstances (e.g., whether we like or dislike the other person). The only force that could make us do morally right things is a supernatural observer who is watching us from Wonderland and who demands that we follow the Golden Rule. In modern monotheistic religions, the holder of morality is God. Although consciously a person may consider themselves an atheist, subconsciously they may still believe that God exists.

The New Testament says, "You have heard that it was said, 'You shall not commit adultery.' But I tell you that anyone who looks at a woman lustfully has already committed adultery with her in his heart" (Matthew, 5:27–28). For believers in God, "committing adultery in his heart" means the choice of evil, and this choice is punishable. But what about non-believers? It appears that all people who choose morally good actions have to be believers in God. Of course, they may not realise they are believers, but they are. They are because they behave as if they believe that there is an observer in Wonderland who is watching their actions, like a CCTV camera of some kind. This 'practical belief' can be accompanied by the absence of any conscious belief in God. Some psychological experiments suggest that the proportion of such people is about 20% of the population[25]. I think that people choose evil only if their subconscious belief in God is too weak to overcome their natural egotism. Evil has many faces – from a commonplace disregard of other people's interests to murder – and there are many ways of justifying doing evil things. But there is only one way to justify doing good things if doing good things is unprofitable and unobserved by others – the belief in a supernatural 'enforcer'. The role of such a belief becomes especially evident in extreme circumstances in which all hopes of being rewarded for doing good vanish. In such circumstances, the 'personal myths' that people create about themselves disappear and people's souls are stripped naked. Russian writer and scientist Dmitry Likhachov, who survived the Leningrad blockade by the German army in 1941–1944, writes, "Starvation is incompatible with any reality, with anything we know of life where food is available … I believe that the real life is starvation; any other life is a mirage. During starvation people showed who they really were, their souls stripped naked, got free of all the trumpery: some proved to be wonderful, unparalleled heroes, and the others – villains, scumbags, murderers, cannibals … everything was genuine. The heavens burst open, and God could be seen in the heavens. He was clearly seen by good people. Miracles were happening"[26], p. 369, translated from Russain by the author].

According to the Old Testament, paradise was Wonderland. God talked to people, lions roamed peacefully next to sheep and people were immortal and didn't know evil. Expelled from paradise, people were thrown into the world of the 'laws of nature': they had to plough, sow, shepherd cattle and 'eat bread by the sweat of their brows'. And they met evil. Evil told a person, "You are a master of this world. Other people are nothing but feeble copies of yourself on the screen of your mind. Do always as you please if the circumstances are right." And only those who retained their memory of Wonderland were able to withstand evil's charm. *Without modern people's explicit or implicit belief in God and the supernatural, universal morality is nothing but an abstract theory or a self-serving illusion at best.*

The magic of the brain

The title of a recent book by a prominent Russian physiologist Natalia Bekhtereva, *The Magic of the Brain and the Labyrinths of Life*[27], reflects the popular attitude towards the brain as a magical entity that governs a human person. Of course, calling the brain's work 'magical' is only a metaphor; scientists don't allow for any magic to enter the work of the brain. But a person's life is more than the person's profession. Somehow, scientists who study the brain are not indifferent to magic. It appears that something that we usually call natural suddenly becomes supernatural, just as we slightly change our angle of view. For example, a complex biological computer – the brain – all of a sudden and due to unknown causes becomes illuminated by the magical light of subjective experience – by consciousness, sensations and desires.

Indeed, why doesn't a person feel pain when he or she is under general anaesthetic? The mechanism that turns a distortion in the body into the feeling of pain is switched off. In other words, the link between the body and Wonderland is temporarily suspended. Indeed, as argued in Chapter 3, subjective experience of a simple sensation is unexplainable in terms of physics and physiology. Usually, we are unaware of this because the 'magic of sensation' is such a commonplace event. Of course, some theorists declare subjective experiences – pain, love, attention, free will and creative insight – to be mere illusions[28,29]. In such theorists' views, the brain is nothing but a complex machine. But here some problems arise. First, it is not clear why the brain needs the illusion of subjective experiences in the first place. If sensations such as pain and pleasure are useless, then why are they there? And if they are useful, they are not illusions. Second, it is not clear how the authors of the 'brain machine' theory can even understand the term 'subjective experiences', because their own brains have to be machines as well, and a machine cannot invent the idea of subjective experiences. It appears that the authors of the 'brain machine' theory make an exception for their own brains and view their brains as holders of something non-mechanical and immaterial, something that we call subjective experience. In reality, the authors are looking at their own brains from the Wonderland of their subjective experience, which they declare to be an illusion. Viewing our own brains from Wonderland is a very comfortable position, since our brains are in front us in full view as a phenomenon and as a scientific concept, allowing all sorts of studies to be done with them. But then we have to acknowledge that Wonderland is not a fiction, but a fact, and a useful fact at that.

Some scientists go all the way, from initially denying the reality of Wonderland to acknowledging in the end that Wonderland really exists. For example, neurobiologists Newberg, d'Aquili and Rause started their investigation into mystical experiences by considering the magical world to be an illusion produced by the brain. According to these authors, God exists as a concept or as a reality only in a person's mind[30]. This point of view suggests that the brain produces the concept of God like the liver produces bile. The authors even found the area in the brain that is responsible for the altered states of consciousness known as 'mystical experiences'. However, by the end of their investigation, the authors had to acknowledge that the higher reality described by mystics exists independently of the brain and is even more important than physical reality. The evolution of their views brought the authors to the conclusion that the brain does not create the mystical reality; rather, the brain is like an eye capable of perceiving the mystical reality. In the aforementioned book, Natalia Bekhtereva also writes about magical phenomena such as clairvoyance and prophetic dreams.

So, why do some physiologists and biologists use the term 'magic' when they describe the brain's work? Do they do this in order to metaphorically express the view that our knowledge of the brain is limited? Or perhaps *they do this to acknowledge that the brain is not only the 'seat of consciousness', but also the organ that links us to the magical reality.*

Magic in human relationships

A person who is in love with another person knows the secret that is unknown to those who have never been in love. The secret is that the object of love – willingly or not – acquires power over the person who is in love, because the loved one can decide on his or her whim whether to satisfy or not satisfy the loving person's passionate desire. The loved one becomes a part of the person who is in love, and sometimes this part is a more valuable part than the loving person is to himself or herself. In the person who is in love, the instinct of self-preservations is suppressed, and the person may sacrifice for the loved one the most valuable parts of himself or herself – career, honour, property and sometimes life itself.

Literary examples of this are well known. Romeo and Juliet committed suicide when they realised that their loved one was dead. The character of the novella by Stefan Zweig *Amok* gave up his career and life with the aim of saving the reputation of a woman he had suddenly fallen in love with. The statistics of suicides based on unshared love are a sad confirmation of these literary examples. Compared to other examples of suicidal behaviour (e.g., psychotic conditions due to drug abuse), the number of love-based suicides is relatively small, yet notable[31]. Out of 30 listed causes of suicidal behaviour, two belong to the death of a loved one and divorce[32].

But is sexual love of another person the only form of love? And how about the love of God, of our children, of our profession or just of certain things? Indeed, the object of love acquires power over the loving person independently of who or what this object is, and for the sake of having access to the object of love, the person who is in love is ready to give up his or her most vital needs. Suicidal terrorists blow themselves up with the words of their love for God, collectors are ready to pay fortunes for the sake of obtaining the object of their passion and a person's love of his or her profession can make the person's life impoverished in other respects. The character of another short story by Stefan Zweig, Buchmendel, is an eccentric but brilliant book peddler who was so deeply captured by his love of his profession that he simply failed to notice that war had broken out; as a result, he lost his freedom and ultimately his life. Unlike love, other feelings, such as attachment, sympathy, interest and curiosity, attract us to objects but don't ruin the basic hierarchy of our motives and stay under our rational control.

So, what makes a person in love go to such great lengths in order to get in close contact with the object that, from a rational point of view, the person might be able to live without? *It can be assumed that the force that drives the person in love to the object of love is magical participation.* As mentioned in previous chapters, the person who is engaged in participation unwittingly crosses the border that separates mental processes (e.g., the feeling of disgust elicited by seeing a piece of chocolate shaped into the form of dog faeces) from real objects (e.g., the piece of chocolate that is entirely suitable for consumption). When falling in love, the person's subconscious choices get fixed on a certain object with which the person identifies himself or herself in part or completely. The factors that influence this identification are so complex and bizarre that they can hardly be a subject of systematic analysis. Phenomenological and artistic descriptions are better suited for such an analysis. Speaking metaphorically, love is the trunk of a tree whose roots, dividing and branching in the most unpredictable ways, go deep into the biological and cultural underground of a human individual. Still, there is something to 'being in love' that cannot be reduced to natural causes, and this something is the fact that *the object of love is not a part of the external world, but a part of our own mind.* We are in love not with a real person 'out there', but with the image of that person we created in our mind. *Love is therefore a magical phenomenon – an interaction between two processes within our mind: the need to fall in love and the image of the person we choose to fall in love with* (see Table 0.3).

It is not a coincidence that attempts to attract love by magical means have a very long history[33]. Love magic was widely practiced in ancient Egypt, Greece and Rome and among the Gallic and Celtic tribes of Europe and Britain[34]. One of the first motives on this theme is the myth of Hercules' wife Deianira, who tried to win her frivolous husband's faithfulness by smearing his cloak with a magical potion given to her by the treacherous centaur Nessus, unwittingly bringing Hercules to his death. In Europe of the Renaissance (fourteenth to seventeenth centuries), love magic was often interweaved with Christian ritual routine: people used to hide clay dolls or rolls with love spells in churches, and church candles were lit during magical love rites[35]. Interestingly, in the time of the Renaissance, love magic was predominantly associated with women and included such aspects of female sexuality as fertility, menstruation and the genitals, whereas men were viewed as unrelated to love magic. This one-sided attribution of love magic to women was evident in the trials of the Holy Inquisition: most of these trials were of women accused of bewitching men. Thus, the medieval text *Malleus Maleficarum* (*The Hammer of Witches*), written in 1487 by the German catholic priest Heinrich Kramer, attributes women's inclination for love magic to their insatiable passion instilled in them by the devil[36]. The association of love magic with women could, however, have been a bias caused by the fact that the majority of the authors writing on love magic were male. There are also data testifying that men applied love magic as frequently as women; for instance, in order to attract the love of a desirable but unreachable woman[37]. Today, like in ancient times, love magic occupies a noticeable position in life. A search using Google for 'witchcraft and love' brings over 27 million references, and for 'love magic' brings over 1,200,000. Egyptian love spells are widely used[38], followed by Wiccan love-attracting spells[39] and others.

In the English-speaking world today, it is common to use the term 'chemistry' when talking about love, meaning that a person in love gets united with the object of love like one chemical element with the other, thus creating a novel entity. This deceptively simple metaphor is quite exact, as it captures the magical nature of love based on the rationally unexplainable phenomenon of the change of quality. Indeed, it is impossible to explain in rational terms why the unification of two elements, hydrogen and oxygen, which under normal conditions are gases, results in an entirely new substance – water, which possesses properties different from those of the gases. Chemistry and physics can explain the mechanism of the unification, but not why the unification leads to the emergence of new qualities. The new qualities in complex chemical substances cannot be reduced to or deduced from the qualities of the simpler substances. The emergence of new qualities is a phenomenon akin to the phenomenon of magical emergence/vanishing (see Table 4.1).

Still, in cognitive science today, attempts have been made to link the phenomenon of love to biochemical processes in the brain. For instance, the specialist on neuroethics Brian Earp has suggested that love is a phenomenon rooted in ancient neurochemical systems that evolved for supporting the reproductive activities of organisms in our ancestors[40]. According to this scientist, in some respects, the neurochemical basis of love is similar to the neurochemical mechanisms of drug addiction. Like drug addiction, love can usurp the person's will and deprive the person of the ability to make rational decisions. This opens up the possibility of 'curing' a person of love by means of giving them special 'love-reducing' pharmacological substances. In an interview, Earp acknowledged that our experience of being in love couldn't be reduced to brain chemistry, though he did not clarify what exactly distinguishes the experience of love from its neurochemical basis. Yet this distinction is clear enough: it is impossible to logically deduce the experience of being in love from brain processes in the same way that it is impossible to deduce the physical and chemical properties of water from the properties of hydrogen and oxygen. *The experience of being in love is a novel quality that obeys not the laws of brain chemistry, but the law of magical thinking – participation.*

So, is it really possible to cure a person of the magic of love? Research has shown that an increase in blood levels of serotonin – a mood-regulating hormone – can reduce the suffering of patients with obsessive–compulsive disorder. A similar increase of serotonin blood levels was observed in people when their love passion was fading. This correlation allowed some scientists to suggest that increasing serotonin blood levels might reduce a person's love passion[41]. Drugs that do this are called antidepressants. But using antidepressants for curing the experience of love is the same as trying to fix a TV set by punching it. Antidepressants reduce the general level of the human ability to experience romantic attachment to people. The fact of converting a person in love into an emotionally insensitive individual says nothing about the unique passion that love represents. What this fact confirms is that the intrusion into brain processes can influence subjective experiences, but the laws these experiences obey remain independent of brain processes. Clearly, if a piano isn't properly tuned, then the reproduction of a musical composition played on this piano could suffer, but the magical harmony of the musical composition in its own right is independent of the condition of the piano's bars and strings.

Getting into Wonderland

Historically, people used various ways of accessing Wonderland. In ancient times and in the Middle Ages, religion, witchcraft, astrology and alchemy provided such access. In the time of the Renaissance, art – paintings, poetry and literature – joined in. In the nineteenth century, the spiritualist movement appeared, and in the twentieth century – parapsychology. Cinema and TV provided new opportunities for the visual representation of Wonderland in the form of such masterpieces as the films by Walt Disney and Alfred Hitchcock or, more recently, the *Lord of the Rings*, *Narnia* and *Harry Potter* series. Finally, at the end of the twentieth and beginning of the twenty-first centuries, there appeared interactive forms of visual representation of Wonderland – computer games and the Internet. Electronic gadgets (notebooks, androids and iPhones) brought Wonderland into the family and children's rooms. Excursions into Wonderland became routine and easily affordable.

Simultaneously with the growing accessibility of Wonderland, another process was going on: the fading of traditional religious feelings in the modern urban population. In the previous centuries, the belief in God connected people with the supernatural. In the modern industrial world, people began to increasingly pay attention to the material side of life. Technology, medicine and education made the life of a modern Western individual safer and more comfortable, but they didn't make the individual free of worries, frustrations, illnesses and death. Technological progress cannot teach us what to do in situations of moral choices, nor answer the question of what meaning our lives have beyond the simple perpetuation of existence. But most importantly, science and technology did not change the essence of a human being – the need for a person to feel that he or she came into this world for something more valuable than mere survival. An existential vacuum emerged in the hearts of modern people. When a person is healthy, not hungry and has a shelter, that person begins to feel bored, and there appears the urge to peep into Wonderland. Some people still seek Wonderland in church, but the number of such people is plummeting. And what do those people do who have lost their belief in God, or never acquired this belief in the first place?

The simplest way to get in touch with Wonderland is using substances like alcohol or cocaine, but this is damaging and dangerous. It is much better to go to a nice concert, read a good book or let oneself be carried away by a daydream, but for this one needs education and a powerful imagination – the skills of understanding music, pondering over a book and the ability to dream. A more accessible way is to 'buy a dream' by plunging into the virtual reality of cinema or computer games. By identifying themselves with movie characters, people can experience the illusion of their own value and power. By plunging into Wonderland, we temporarily escape from the monotonous predictability of everyday life. In this way, *magical thinking fills the existential vacuum that torments many people today*. The entertainment and computer game industries exploit the need of modern people for accessible ways of getting in touch with magical reality by stuffing the market with movies and electronic games. There also exists a more positive trend in exploiting people's magical thinking – using magical thinking for psychotherapy and the development of the imagination[42]. Yet, as recent studies in psychology have shown, the penetration of Wonderland into the lives of modern people goes beyond magical thinking alone.

Rapa Nui: conclusion

This chapter began with the following questions: why do rational people like to play with magical reality? What psychological consequences can this fascination with magical reality bring about?

Experiments have revealed that in modern educated adults, the belief in the supernatural did not cease to exist. Under the pressure of science, this belief descended into the subconscious, but continues to filter into various domains of modern life: economics, medicine, morality, art, politics, education and the theories of modern physics and astronomy. This hidden belief in the supernatural can explain the phenomena of modern life that would be hard to explain without such belief. Why do rational people, when they are faced with choices in economics, often follow the laws of magic rather than common logic? What makes modern educated individuals follow political ideas that contradict their conscious interests? Why does the placebo effect exist? From where do suicidal terrorists take their courage to commit actions of self-destruction? How is it possible that some people make moral choices and sacrifice their private interests even when there is no surveillance? Why do some scientists call the work of the brain magical? Why is being in love called chemistry? How is it possible that the whole universe once had a volume smaller than a grain of sand?

Excursions into Wonderland are not for entertainment only; we need them to fill the existential gap in our souls. Our tendency to engage with the supernatural, sometimes at great cost, reminds me of the Rapa Nui people who lived on Easter Island, lost in the vastness of the Pacific. The history of this people is sometimes quoted as an example of a wasteful attitude towards natural resources. For the sake of the transportation and erection of giant stone statues called Moai, the Rapa Nui people completely destroyed their island's luxurious palm groves; without the building materials for rafts and canoes, the islanders were unable to fish and their culture fell into decay.

While this may be true, what would this small piece of land be without its Moai statues? It would remain an insignificant little island, one of thousands scattered around the ocean. The Rapa Nui people would be nothing to talk about apart from them being an ethnic group that speaks a Polynesian language. The whole greatness of the Rapa Nui people is in the fact that these people did not simply believe in their gods, but made a creative effort to build the giant statues that impersonated the gods. In terms of the physical survival of the people, this creative effort was meaningless and self-destructive, but its cultural significance was immense. The creators of the Moai can be compared with the anonymous artists of the Upper Palaeolithic who covered their caves with magnificent paintings of animals and people, and with the builders of Gothic cathedrals.

Perhaps all that we call the great achievements of culture is exactly that – a self-destructive super-effort that people make against the voice of reason and the instinct of self-preservation due to the unstoppable urge to get in touch with Wonderland. The Rapa Nui people were unfortunate, since they lacked external sources for restoring the damage caused to their environment by their artistic effort. The modern descendants of this people live a very ordinary life and don't resemble their famous ancestors in any way. Today, the nations of the West have stepped on the Rapa Nui ancestors' path. Billions of dollars are spent on the construction of machines that accelerate elementary particles of matter almost to the speed of light, but don't have much utilitarian value. The same can be said of the giant telescopes that can see the remote edges of the visible universe. A final product of these supercolliders and mega-telescopes is the real Wonderland – the concepts and theories that disturb the otherwise consistent and unified temple of modern science. These concepts and theories include the Big Bang, quantum entanglement, parallel universes and other miracles of modern theoretical physics.

Magic is incomprehensible, potentially dangerous and rejected by science and religion, yet it remains irresistibly attractive to people. Experiments have shown that if a person is given a choice of seeing either 'real magic' or a new and exciting scientific effect, most people (children and adults alike) choose 'real magic'[43]. For children, this choice is understandable, but why do most adults make this choice? Could it not be because, at the bottom of their hearts, every person still harbours a belief that the world of the supernatural is not a dream, but reality? Against all odds – the theories of science, the efforts of school education and the testimony of everyday experience – modern people still believe in miracles. Because if Wonderland is real, that might be something. Perhaps, that could even make us immortal.

Notes

1 http://mashable.com/2017/12/28/gaming-disorder/#.1Lf4drBEaqE
2 http://ru.wikipedia.org/wiki/Spherical..geometry
3 Nemeroff, C. & Rozin, P. (2000)
4 Subbotsky, E. (2014)
5 Lee, C., Linkenauger, S. A., Bakdash, J. Z., Joy-Gaba, J. A. & Profitt, D. R. (2011)
6 Subbotsky, E. (2005)
7 Subbotsky, E. (2007)
8 Kahneman, D. & Tversky, A. (1973)
9 Tversky, A. & Kahneman, D. (1974)
10 Rozin, P., Millman, L. & Nemeroff, C. (1986)
11 Galbraith, J. K. (2009)
12 Yang, M. & Roskos-Ewoldsen, D. R. (2007)
13 Shu, S. B. & Carlson, K. A. (2014)
14 http://en.wikipedia.org/wiki/3_(number)
15 Subbotsky, E. & Matthews, J. (2011)
16 Hróbjartsson, A. & Norup M. (2003)
17 Kaptchuk, T. J., Friedlander, E., Kelley, J. M., et al. (2010)
18 Brooks, M. (2008)
19 http://en.wikipedia.org/wiki/Homeopathy
20 Miller, D. (2011)
21 http://ru.wikipedia.org/wiki/Хирохито
22 Ivanov, U. G. (2001)

23 Atran, S. (2003)
24 Kara-Murza, S. G. (2007)
25 Batson, C. D. & Thompson, E. R. (2001)
26 Likhachev, D. (2013)
27 Bekhtereva, N. (1999)
28 Dawkins, R. (1976)
29 Dennett, D. (1991)
30 Newberg, A., d'Aquili, E. & Rause, V. (2013)
31 http://psihomed.com/suitsid/
32 www.suicide.org/suicide-causes.html
33 Rosenblatt, P. C. (1971)
34 https://en.wikipedia.org/wiki/Love_magic
35 Ruggiero, G. (1993)
36 Holdrige, B. (1974)
37 Dickie, M. W. (2000)
38 www.egyptian-witchcraft.com
39 http://wiccanspells.info/category/love-spells/
40 Robson, D. (2014)
41 Thomson, H. (2014)
42 Matthews, J. & Matthews, C. (2009)
43 Subbotsky, E. (2010a)

EPILOGUE: IMAGINING THE UNIMAGINABLE

Science liberates, but it also hypnotises: it makes us insensitive to the reality of the supernatural. In order to free ourselves from this hypnotic dream, let's conduct a thought experiment. In this experiment, we imagine that magic and magical thinking have disappeared from the world. What kind of world would that be?

Of course, imagining a world like that is a contradiction in terms, because imagination itself is a part of subjective experience, and subjective experience by definition is a supernatural phenomenon. We know that galaxies, stars, planets and the whole universe exist in their present shape because we know about them: we see them through telescopes, develop theories about them, model them in our computers and write books and papers about them, and all of this is based on subjective experience. With subjective experience gone, the physical universe would of course remain, but we wouldn't be able to say what this 'unconscious' universe is like. So, it looks like we are trying to imagine the unimaginable.

But there is a way out of this paradox. We can try and imagine the magic-free universe 'as if' we were still seeing this universe through some kind of bodiless 'god's eye'. After all, we can imagine the universe as it existed for billions of years before life appeared on earth. Of course, we know that this early universe is still a product of our subjective activities: we reconstruct the 'universe back then' on the grounds of the universe we see now, but this reconstructed universe is still a useful model that helps us understand the world. Like the 'impossible waterfall' in the lithograph of Maurits Escher, where water seems to flow upwards, the image of the 'unconscious universe' is an illusion, yet it appears real. So, let's pretend that we see a world in which there are no supernatural phenomena, but there are still intelligent beings.

What kind of beings would they be? Certainly, they wouldn't be living creatures, because life, too, is a supernatural phenomenon. Rather, these creatures would be similar to intelligent machines. A creature like that would have intelligence, but its intelligence would be different from what we call human thinking. Rather, a more suitable term for this intelligence would be 'information processing'. The difference between human thinking and information processing is that thinking is a form of subjective experience – a part of the human soul, and information processing doesn't need a soul. A soul, or subjective experience, is the magical entity that the creatures of our utopian world would lack. So, let us call these creatures 'nosouls'. Let us imagine what nosouls could and couldn't do.

To begin with, nosouls wouldn't be able to see dreams or experience feelings through immersing themselves in the magical world of art or taking hallucinogenic drugs. For these intelligent creatures, love would lose its 'chemistry' – the irrational element that is based on magical participation – and their relationships would be reduced to marital contracts and prenup agreements. There would be no moral rules sanctioned by gods, only social contracts sanctioned by positive or negative reinforcement (e.g., switching the energy supply off for disobedience). There would be neither fantastic worlds of the imagination, nor parallel invisible worlds in which gods and the spirits of dead ancestors live. The language of nosouls would be similar to the digital languages of computer programming, incapable of carrying metaphorical meanings and animistic expressions, such as 'a rising sun' or 'a flying arrow'.

And what about art? If we are to believe that human art emerged as a way of communicating with ancestors' spirits, nosouls wouldn't have this kind of art. Their art might emerge as the art of design; this art would be strictly rational, like computer graphics or sculptures created by machines. Trends in art would disappear. Art similar to suprematist paintings by Kazimir Malevich and abstract geometric fantasies by Wassily Kandinsky might exist, but the surrealism of Giorgio de Chirico and René Magritte would not. Nosouls would be able to create realistic stories in the spirit of some Charles Dickens' novels, but the magical realism of Alejo Carpentier and Jorge Luis Borges would be unknown to them. Repetitive rhythmic music, however complex, would stay, but the magically inspired, unpredictable melodies of Mozart or Tchaikovsky would be impossible.

Religion? There would be none. Nosouls wouldn't fear death, and the thought of the afterlife would never enter their digital brains. Without myth and religion, there would be no archetypical stories able to inspire poetry and art. Luckily for the nosouls, religious wars and suicidal terror would also have no grounds. It is hard to imagine nosouls searching for the meaning of life. Perhaps, like in most animals, having a meaning to life simply wouldn't be necessary to nosouls – only the endless perpetuation of themselves, both in time and space.

Nosouls could probably develop science, but they would not be able to experience leaps of creative imagination. The soulless 'Einstein' would never fly on the edge of a light beam, and the soulless 'Maxwell' could never imagine a demon sorting molecules into slow and fast ones. Human scientists receive inspiration from the creative subconscious combinations of ideas produced by magical participation, but nosouls wouldn't have the subconscious; instead, they would have to logically induce new combinations from the already available ones, and this would make their scientific progress very slow.

For psychology, the loss of magical thinking would be devastating. With no subjective experience, psychology would lose its subject and turn into the science of catching and fixing computer bugs. Some modern physicists and psychologists would be happy to learn that in the nosouls' world, parapsychology would be impossible. Being a machine or a puppet in a simulated cosmic computer, nosouls would be completely merged with the stuff of the surrounding universe, and the issue of affecting this stuff with the efforts of their digital mind could never arise. But nosouls' physical science would also suffer. Without subjective experience, there would be no fundamental opposition between an observer and the observed; instead, reality would simply leave its imprints in the nosouls' minds, like we leave the imprints of our feet in wet sand. It would therefore never occur to nosouls to conduct the double-slit experiment and discover the effect of the observer on the observed.

On the positive side, in the utopian world of nosouls, the losses would be partially compensated by some gains. Nosouls would be free from the temptation of getting immersed in the magical world of drug-induced hallucinations, and mental disorders such as schizophrenia or obsessive–compulsive disorder would be unheard of. There would be no Holy Inquisition, witch-hunting or religious wars. Manipulation of the mass consciousness on the basis of a belief in the supernatural would be impossible, but authorities would have to control their subjects by sheer force. Mass media would be free from charlatans and 'specialists' with paranormal abilities. Perhaps nosouls would be immortal and able to live in space. Feeding on cosmic radiation, they could probably leave earth and travel in the universe between the stars and galaxies.

Would they have a childhood? If they did, that childhood must be a dull one. Without magical Wonderland, without wizards and fairies, the world of nosouls' children would have to be reduced to the accumulation of information and learning of artificial languages. That said, nosouls' kids would probably be outstanding chess players from the age of nursery and great at maths at secondary school.

With all these losses and gains, a utopian world without magic would have little in common with the world we live in. Because the world of nosouls is impossible, we don't have to be worried about the problems of this fantastical world. Yet imagining the unimaginable helps us appreciate the role of magic in modern life. The most important role of magical thinking is that it is the ground on which the whole edifice of our mind is built. Our emotional reactions and communicative interactions are based on sympathetic magic. Psychological effects such as emotional contamination, hypnotic suggestion or the placebo effect are just a few examples of magic-based communicative interactions. In love and in child caring we often apply magical rituals, such as hugging, giving presents or making compliments, which from the rational point of view don't make much sense. But these rituals are a vital part of what poets call 'the magic of love' and lovers call 'chemistry'. Whereas rational and scientific thinking helps us cope with the problems of the physical world, magical thinking and a belief in magic and in God come to our aid when we struggle with problems in our social and emotional lives. This is why magical thinking and a belief in magic (including traditional religious beliefs) do not contradict logic and science, but complement them, giving our lives excitement and meaning.

Today, a popular topic of discussion is the relationship between people and machines. According to some prophets, in the future, a human being will become a cyborg – the merger between a person and a computer. Other theorists predict that artificial intelligence (AI) will progress to the extent that machines become smarter than people and might even take over from humanity. I don't believe in such prognoses. The reason for my disbelief in this dystopian future is that the world we live in is based on magic. It is the magic of human subjective experience that gave birth to rational and scientific thinking, including AI. Rather, I believe that, in the future, the opposite process will take place: people will increasingly rely on the hidden powers of the mind. When rational and scientific thinking reaches the limits of its explanatory capacity, people will have no choice but to return, at a new level, to the ancient belief that the human mind and the universe are inherently linked with one another. The signs of this return are visible even now. In physical science, these signs are theories such as the 'Copenhagen interpretation' and the 'many-worlds interpretation' of quantum events, in cosmology they are theories such as the 'anthropic principle' and in psychology these signs are the studies of psychic phenomena, such as extra-sensory perception or psychokinesis. And although predicting the future is a risky business, I believe that the future world will be a lot less 'inanimate' than the world of today. That is why, along with the progress of scientific exploration, magic and magical thinking are, and will increasingly be, exciting topics for interdisciplinary research.

BIBLIOGRAPHY (TRANSLITERATED)

Achenbach, J. (2007). Programmed for love. *The Washington Post*, 12, 23.

Atkin, C. (1983). Effects of realistic TV violence vs. fictional violence on aggression. *Journalism Quarterly*, 60, 615–621.

Atran, S. (2003). Genesis of suicide terrorism. *Science*, 299, 1534–1539.

Baggott, J. (2013). *Farewell to Reality: How Fairytale Physics Betrays the Search for Scientific Truth*. London: Constable.

Baillargeon, R. (1987). Object permanence in 3 1/2- and 4 1/2-month-old infants. *Developmental Psychology*, 23, 655–664.

Barrow, J. & Tipler, F. (1986). *The Anthropic Cosmological Principle*. New York: Oxford University Press.

Batson, C. D. & Thompson, E. R. (2001). Why don't moral people act morally? Motivational considerations. *Current Directions in Psychological Science*, 10, 54–57.

Bekhtereva, N. (1999). *Magiya mozga i labirinty zhizni*. Saint Petersburg: Notabene. www.e-reading.ws/book.php?book=69185

Bem, D. (2011). Feeling the future: experimental evidence for anomalous retroactive influences on cognition and affect. *Journal of Personality & Social Psychology*, 100, 207–225.

Bem, D. J. & Honorton, C. (1994). Does Psi exist? Replicable evidence for an anomalous process of information transfer. *Psychological Bulletin*, 115, 4–18.

Benton, D., Parker, P. Y. & Donohoe, R. T. (1996). The supply of glucose to the brain and cognitive functioning. *Journal of Biosocial Science*, 28, 463–479.

Berne, J. & Radunsky, V. (2013). *On a Beam of Light: A Story of Albert Einstein*. New York: Chrinicles Books.

Bettelheim, B. (1977). *The Uses of Enchantment: The Meaning and Importance of Fairy Tales*. New York: Vintage Books.

Bloom, P. & Weisberg, D. S. (2007). Childhood origins of aduld resistance to science. *Science*, 316, 996–997.

Bolton, D., Dearsley, P., Madronal-Luque, R. & Baron-Cohen, S. (2002). Magical thinking in childhood and adolescence: development and relation to obsessive compulsion. *British Journal of Developmental Psychology*, 20, 479–494.

Boyer P. (1994). *The Naturalness of Religious Ideas: A Cognitive Theory of Religion*. Berkeley: University of California Press.

Boyer, P. (2001). *Religion Explained: The Human Instincts that Fashion Gods, Spirits and Ancestors.* New York: Basic Books.

Briggs, R. (2012). *Fungus the Bogeyman.* Singapore: Penguin Books.

Bushman, B. J. & Huesmann, L. R. (2001). Effects of televised violence on aggression. In D. G. Singer & J. L. Singer (Eds), *Handbook of Children and the Media.* Thousand Oaks: Sage, pp. 223–254.

Brooks, M. (2008). *13 Things that Don't Make Sense.* New York: Random House.

Cameron, J. & Pierce, W. D. (1994). Reinforcement, reward, and intrinsic motivation: a meta-analysis. *Review of Educational Research,* 64, 363–423.

Carruthers, P. (2011). *The Opacity of Mind: An Integrative Theory of Self-Knowledge.* Oxford: Oxford University Press.

Carter, B. (1974). Large number of coincidences and the anthropic principle in cosmology. In *IAU Symposium 63: Confrontation of Cosmological Theories with Observational Data.* Dordrecht: Reidel, pp. 291–298.

Caspi, A., Sugden, K., Moffitt, T. E., et al. (2003). Influence of life stress on depression: moderation by a polimorphism in the 5-HTT gene. *Science,* 301, 386–389.

Castiglioni, A. (1946). *Adventures of the Mind.* New York: Alfred A. Knopf.

Cialdini, R. B. (2007). *Influence: The Psychology of Persuasion.* New York: HarperCollins.

Clarke, A. C. (1962). Hazards of prophecy: the failure of imagination. In A. C. Clarke (Ed.), *Profiles of the Future: An Enquiry Into the Limits of the Possible. A Collection of Works.* New York: Harper & Row, pp. 19–26.

Cole, M. & Subbotsky, E. (1993). The fate of stages past: reflections on the heterogeneity of thinking from the perspective of cultural–historical psychology. *Schweizerische Zeitschrift fur Psychologie,* 52, 103–113.

Coriat, I. H. (1923). Suggestion as a form of medical magic. *Journal of Abnormal and Social Psychology,* 18, 258–268.

Crick, F. (1994). *The Astonishing Hypothesis: The Scientific Search for the Soul.* London: Simon & Schuster.

Davies, P. (1992). *The Mind of God: The Scientific Basis for a Rational World.* New York: Orion Production.

Davies, P. (2006). *The Goldilocks Enigma: Why is the Universe Just Right for Life?* London: Allen Lane.

Dawkins, R. (1976). *The Selfish Gene.* Oxford: Oxford University Press.

Dawkins, R. (1986). *The Blind Watchmaker.* New York: W. W. Norton & Company.

Dawkins, R. (2012). *The Magic of Reality: How We Know What's Really True.* London: Black Swan.

Day, J. M. (1994). *Plato's Meno in Focus.* London: Routledge.

de Benua, A. (2013). *Kak Mozhno Byt Yazychnikom.* Moscow: Russkaya Pravda.

de Beauregard, O. C., Mattuck, R. D., Josephson, B. D. & Walker, E. H. (1980). Parapsychology: An Exchange. The New York Review of Books. www.nybooks.com/articles/1980/06/26/parapsychology-an-exchange/

de Chardin, P. T. (1961). *The Phenomenon of Man.* San Francisco: Harper & Row.

De Lange, C. (2014). Cure for love: fall for a robot to fend off heartache. *New Scientist,* February 14.

Dennett, D. (1991). *Consciousness Explained.* New York: The Penguin Press.

Dennett, D. C. & Kinsbourne, M. (1992). Time and the observer: the where and when of consciousness in the brain. *Behavioral and Brain Sciences,* 15, 183–247.

DeWitt, B. S. & Graham, R. N. (Eds) (1973). *The Many-Worlds Interpretation of Quantum Mechanics (Princeton Series in Physics).* Princeton: Princeton University Press.

Dias, M. G. & Harris, P. L. (1988). The effect of make-believe play on deductive reasoning. *British Journal of Developmental Psychology,* 6, 207–221.

Dickie, M. W. (2000). Who practiced love-magic in classical antiquity and in the late Roman world? *The Classical Quarterly*, 50, 563–583.

Dolgova, A. I. (Ed.) (2001). *Kriminilogogiya. Uchebnik Dlia Vusov*. Moscow: NORMA Publishing. www.e-reading.club/book.php?book=1001912

Donati, O., Missiroli, G. F. & Pozzi, G. (1973). An experiment on electron interference. *American Journal of Physics*, 41, 639–644.

Dunbar, R. (2014). *Human Evolution*. London: Pelican Books.

Dunne, B. J., Nelson, R. D. & Jahn, R. G. (1989). Operator-related anomalies in a random mechanical cascade. *Journal of Scientific Exploration*, 2, 155–179.

Eagleman, D. (2012). *Incognito*. Edinburgh: Canongate Books.

Eberhard, P. H. & Ross, R. R. (1989). Quantum field theory cannot provide faster-than-light communication. *Foundations of Physics Letters*, 2, 127–149.

Einstein, A. (1934). Prinzipien der Forschung: Rede zum 60. Geburtstag von Max Planck. In *Mein Weltbild*. Amsterdam: Querido, pp. 107–110. English edition: Principles of Research. In C. Seelig (Ed.) (1954), S. Bargmann (trans.), *Ideas and Opinions: Based on Mein Weltbild*. New York: Crown Publishers, 224–227.

Einstein, A., Podolsky, B. & Rosen, N. (1935). Can quantum-mechanical description of physical reality be considered complete? *Physical Review*, 47, 777.

Eliade, M. (1994). *Svyashchennoe i Mirskoe*. Moscow: Izdatel'stvo MGU.

Everett, H. (1957). Relative state formulation of quantum mechanics. *Reviews of Modern Physics*, 29, 454–462.

Eysenck, H. J. (1975). Planets, stars and personality. *New Behaviour*, May 29, 246–249.

Fairclough, S. H. & Houston, K. (2004). A metabolic measure of mental effort. *Biological Psychology*, 66, 177–190.

Farah, M. (2005). Neuroethics: the practical and the philosophical. *Trends in Cognitive Science*, 9, 34–40.

Forgas, J. P. (2002). Feeling and doing: affective influences on interpersonal behavior. *Psychological Inquiry*, 13, 1–28.

Foucault, M. (2003) *The Birth of the Clinic*. London: Routledge.

Frazer, J. G. (1922). *The Golden Bough. A Study in Magic and Religion*. London: Macmillan.

Freud, S. (1935). *A General Introduction to Psychoanalysis*. New York: Liveright.

Gailliot, M. T., Baumeister, R. F., DeWall, C. N., Maner, J. K., Plant, E. A., Tice, D. M., Brewer, L. E. & Schmeichel, B. J. (2007). Self-control relies on glucose as a limited energy source: willpower is more than a metaphor. *Journal of Personality and Social Psychology*, 92, 325–336.

Gailliot, M. T., Baumeister, R. F. & Schmeichel, B. J. (2006) Self-regulatory processes defend against the threat of death: effects of self-control depletion and trait self-conrol on thoughts and fears of dying. *Journal of Personality and Social Psychology*, 91, 49–62.

Galbraith, J. K. (2009). *The Great Crash 1929*. New York: Penguin.

Garcia-Montes, J. M., Peres-Alvarez, M., Balbuena, C. S., Garcelan, S. P. & Cangas, A. J. (2006). Metacognitions in patients with hallucinations and obsessive–compulsive disorder: the superstition factor. *Behavior Research and Therapy*, 44, 1091–1104.

Gasper, K. (2004). Do you see what I see? Affect and visual information processing. *Cognition & Emotion*, 18, 405–421.

Gauquelin, M. (1969). *The Scientific Basis of Astrology*. New York: Stein and Day Publishers.

Gauquelin, M. (1991). *Neo-Astrology: A Copernican Revolution*. London: Arkana, Penguin Group.

Gibson, E. J. & Walker, A. S. (1984). Development of knowledge of visual–tactual affordances of substance. *Child Development*, 55, 453–460.

Gorer, G. (1955). *Exploring English Character*. London: Cresset Press.

Gray, J. (2007). *Black Mass: Apocalyptic Religion and the Death of Utopia*. London: Allen Lane.

Green, B. (2011). *The Hidden Reality: Parallel Universes and the Deep Laws of the Cosmos*. London: Allen Lane.

Gregory, G. H. (1970). *The Intelligent Eye*. London: Weidenfeld & Nicolson.

Gudjonsson, G. H. (1984). A new scale of interrogative suggestibility. *Personality and Individual Differences*, 5, 303–314.

Gudjonsson, G. H. (1987). Historical background to suggestibility: how interrogative suggestibility differs from other types of suggestibility. *Personality and Individual Differences*, 8, 347–355.

Haraldsson, E. (1985). Interrogative suggestibility and its relationship with personality, perceptual defensiveness and extraordinary beliefs. *Personality and Individual Differences*, 6, 765–767.

Hare, B., Call, J. & Tomasello, M. (2006). Chimpanzees deceive a human competitor by hiding. *Cognition*, 101, 495–514.

Harris, P. L., Brown, E., Marriot, C., Whittal, S. & Harmer, S. (1991). Monsters, ghosts and witches: testing the limits of the fantasy–reality distinction in young children. *British Journal of Developmental Psychology*, 9, 105–123.

Hawkins, J., Pea, R. D., Glick, J. & Scribner, S. (1984). Merds that laugh don't like mushrooms: evidence for deductive reasoning by preschoolers. *Developmental Psychology*, 20, 584–594.

Henderson, B. & Moore, S. G. (1980). Children's responses to objects differing in novelty in relation to level of curiosity and adult behaviour. *Child Development*, 51, 457–465.

Hergovich, A. (2003). Field dependence, suggestibility and belief in paranormal phenomena. *Personality and Individual Differences*, 34, 195–209.

Holdrige, B. (1974). *1430–1505 Malleus Maleficarum*. Sound recording by H. Kramer & J. Sprenger; translated by M. Summers. Abridged by B. Holdridge. New York: Caedmon.

Horgan, J. (1997). *The End of Science: Facing the Limits of Knowledge in the Twilight of the Scientific Age*. London: Little, Brown & Co.

Horgan, J. (2004). *Rational Mysticism. Spirituality Meets Science in the Search of Enlightment*. New York: Houghton Mifflin.

Hróbjartsson, A. & Norup, M. (2003). The use of placebo interventions in medical practice – a national questionnaire survey of Danish clinicians. *Evaluation & the Health Professions*, 26, 153–165.

Hsu, C. T., Jacobs, A. M., Altmann, U. & Conrad, M. (2015). The magical activation of left amygdala when reading Harry Potter: An fMRI study on how descriptions of supranatural events entertain and enchant. *PLoS One*, 10, e0118179.

Hutton, R. E. (1999). *The Triumph of the Moon: A History of Modern Pagan Witchcraft*. Oxford and New York: Oxford University Press.

Huxley, A. (1954). *The Doors of Perception*. London: Chatto and Windus.

Ingold, T. (1992). Comment on 'Beyond the original affluent society' by N. Bird-David. *Current Anthropology*, 33, 34–47.

Isaakson, W. (2007). *Einstein: His Life and Universe*. New York: Simon & Shuster.

Ivanov, Yu. G. (2001). *Kamikadze: Piloty-Smertniki. Yaponskoe Samopozhertvovanie vo Vremya Voiny na Tikhom Okeane*. Smolensk: Rusich.

Jahn, R. G. (2001). The challenge of consciousness. *Journal of Scientific Exploration*, 15, 443–457.

Jahn, R. G. & Dunne, B. J. (2008). Change the rules. *Journal of Scientific Exploration*, 22, 193–213.

Jahoda, G. (1969). *The Psychology of Superstition*. London: Penguin.

James, W. (2012). *The Varieties of Religious Experience: A Study in Human Nature*. Oxford: Oxford World's Classics.

Jaynes, J. (1976). *The Origins of Consciousness in the Breakdown of the Bicameral Mind.* New York: Houghton Mifflin.

Johnson, C. & Harris, P. L. (1994). Magic: special but not excluded. *British Journal of Developmental Psychology,* 12, 35–52.

Jung, C. G. (1960). *On the Nature of the Psyche.* Princeton: Princeton University Press (original work published 1928).

Kahneman, D. & Tversky, A. (1973). On the psychology of prediction. *Psychological Review,* 80, 237–251.

Kaku, M. (2008). *Physics of the Impossible. A Scientific Exploration Into the World of Phasers, Force Fields, Teleportation, and Time Travel.* New York: Doubleday.

Kaku, M. (2014). *The Future of the Mind: The Scientific Quest to Understand, Enhance, and Empower the Mind.* New York: Doubleday.

Kandinsky, V. Kh. (2007). *O Psevdogallyutsinatsiyakh.* Moscow: Meditsinskaya Kniga (first published in 1885).

Kaptchuk, T. J., Friedlander, E., Kelley, J. M., et al. (2010). Placebos without deception: a randomized controlled trial in irritable bowel syndrome. *PLoS One,* 5, e15591.

Kara-Murza, S. G. (2007). *Vlast' Manipulyatsii.* Moscow: Akademicheskii proekt.

Karmiloff-Smith, A. (1989). Constraints on representational change: evidence from children's drawings. *Cognition,* 34, 57–83.

Keinan, G. (1994). Effects of stress and tolerance of ambiguity on magical thinking. *Journal of Personality and Social Psychology,* 67, 48–55.

Koberinski, A. & Müller, M. P. (2017). Quantum theory as a principle theory: Insights from an information-theoretic reconstruction. https://arxiv.org/abs/1707.05602

Krupp, E. C. (2003). *Echoes of the Ancient Skies: The Astronomy of Lost Civilizations.* Mineola: Dover Publications.

Kuhn, G., Amlani, A. A., & Rensink, R. A. (2008). Towards a science of magic. *Trends in Cognitive Science,* 9, 349–354.

Langer, E. J. (1975). The illusion of control. *Journal of Personality and Social Psychology,* 32, 311–328.

Lauveng, A. (2015). *Tomorrow I Was Always a Lion.* Samara: Bakhrakh-M.

Leakey, R. (1994). *The Origin of Humankind.* New York: Basic Books.

Lee, C., Linkenauger, S. A., Bakdash, J. Z., Joy-Gaba, J. A. & Profitt, D. R. (2011). Putting like a pro: the role of positive contagion in golf performance and perception. *PLoS One,* 6, e26016.

Leevers, H. J. & Harris, P. L. (1999). Persisting effects of instruction on young children's syllogistic reasoning with incongruent and abstract premises. *Thinking & Reasoning,* 5, 145–173.

Lehrer J. (2010). The truth wears off. *The New Yorker,* 52–57.

Lepper, M. R., Corpus, J. H. & Iyengar, S. S. (2005). Intrinsic and extrinsic motivational orientations in the classroom: age differences and academic correlates. *Journal of Educational Psychology,* 97, 184–196.

Leroi-Gourhan, A. (1968). *The Art of Prehistoric Man in Western Europe.* London: Thames & Hudson.

Levy, D. (2007) *Love and Sex with Robots: The Evolution of Human-Robot Relationships.* New York: Harper

Lévy-Bruhl, L. (1923). *The Primitive Mentality* (English translation). Paris: Alcan.

Lévy-Bruhl, L. (1985/1926). *How Natives Think.* Princeton: Princeton University Press (original work published 1926).

Lewis-Williams, J. D. & Dowson, T. A. (1988). Signs of all times: entopic phenomena in Upper Paleolitic art. *Current Anthropology,* 29, 201–245.

Libet, B. (1999). Do we have free will? *Journal of Consciousness Studies*, 6, 47–57.

Libet, B. (2003). Can conscious experience affect brain activity? *Journal of Consciousness Studies*, 10, 24–28.

Lillard, A. S. (1996). Body or mind: Children's categorizing of pretense. *Child Development*, 67, 1717–1734.

Lillard, A. S. & Sobel, D. (1999). Lion kings or puppies: The influence of fantasy on children's understanding of pretense. *Developmental Science*, 2, 75–80.

Likhachev, D. (2013). *Mysli o Zhizni. Pis'ma o Dobrom*. Moscow: Kolibri.

Lloyd, S. (2007). Wired for romance. *Los Angeles Times*. http://articles.latimes.com/2007/nov/25/books/bk-lloyd25

Luhrman, T. M. (1989). *Persuasions of the Witch's Craft: Ritual Magic and Witchcraft in Present-Day England*. Oxford: Blackwell.

Malinowski, B. (1935). *Coral Gardens and Their Magic*. London: Allen & Unwin.

Mamardashvili, M. & Pyatigorskii, A. (2009). *Simvol i Soznanie: Metafizicheskie Rassuzhdeniya o Soznanii, Simvolike i Yazyke*. Moscow: Progress-Traditsiya.

Marzano, R. J. (2007). *The Art of Science and Teaching*. Alexandria: ASCD.

Marzano, R. J., Marzano, J. S, & Pickering, D. J. (2003). *Classroom Management that Works. Research Based Strategies for Every Teacher*. Alexandria: ASCD.

Matthews, J. & Matthews, C. (2009). *Storytelling Book*. Hechan: The Templar Company.

May, E. C., Utts, J. M. & Spottiswoode, S. J. P. (1995a). Decision augmentation theory: Applications to the random number generator database. *Journal of Scientific Exploration*, 59, 195-220.

May, E. C., Utts, J. M. & Spottiswoode, S. J. P. (1995b). Decision augmentation theory: Applications to the random number generator database. *Journal of Scientific Exploration*, 9, 453–488.

Mayo, J., White, O. & Eysenck, H. J. (1978). An empirical study of the relation between astrological factors and personality. *Journal of Social Psychology*, 105, 229–236.

Mendel, G. (1965). Children's preferences for differing degrees of novelty. *Child Development*, 36, 453–465.

Michelet, J. (1998). *Satanism and Witchcraft: The Classic Study of Medieval Superstition*. London: Kensington Publishing Corporation.

Milgram, S. (1992). *The Individual in a Social World: Essays and Experiments* (2nd ed.). New York: McGraw-Hill.

Miller, D. (2011). *Magiya – v Pomoshch'. Kak Zashchitit'sya ot Magicheskikh I Psikhicheskikh Atak*. Saint Petersburg: Izdatel'skaya Gruppa.

Mindel, A. (1993). *The Shaman's Body: A New Shamanism for Transferring Healthy Relationships and Community*. San Francisco: Harper.

Mithen, S. (2005). *The Prehistory of the Mind. A Search for the Origins of Art, Religion and Science*. London: Thames & Hudson.

Moody, R. (1975) *Life After Life: The Investigation of a Phenomenon – Survival of Bodily Death*. New York: Bantam.

Moravec, H. (1988). *Mind Children: The Future of Robot and Human Intelligence*. Cambridge: Harvard University Press.

Morgan, E. (1982). *The Aquatic Ape*. New York: Stein & Day.

Morrison, P. & Gardner, H. (1978). Dragons and dinosaurs: the child's capacity to differentiate ordinary from fantastic reality. *Child Development*, 49, 642–648.

Nemeroff, C. & Rozin, P. (1992). Sympathetic magical beliefs and kosher dietary practice: the interaction of rules and feelings. *Ethos*, 20, 96–115.

Nemeroff, C. & Rozin, P. (2000). The making of the magical mind: the nature and function of sympathetic magical thinking. In K. S. Rosengren, C. N. Johnson & P. L.

Harris (Eds), *Imagining the Impossible: Magical, Scientific and Religious Thinking in Children*. Cambridge: Cambridge University Press, pp. 1–34.

Nelson, R. D., Bradish, G. J., Jahn, R. G. & Dunne, B. J. (1994). A linear pendulum experiment: effects of operator intention on damping rate. *Journal of Scientific Exploration*, 4, 471–489.

Newberg, A. & d'Aquili, E. (1999). *The Mystical Mind*. Minneapolis: Fortress Press.

Newberg, A., d'Aquili, E. & Rause, V. (2002). *Why God Won't Go Away: Brain Science and the Biology of Belief*. New York: Ballantine Books.

Onishi, K. H. & Baillargeon, R. (2005). Do 15-months-old infants understand false beliefs? *Science*, 308, 255–258.

Ortega y Gasset, J. (1932). *The Revolt of the Masses*. New York: W. W. Norton & Company.

Penfield, W. (1975). *The Mystery of the Mind*. Princeton: Princeton University Press.

Persinger, M. (1987). *Neuropsychological Bases of God Beliefs*. New York: Praeger.

Persinger, M. A. (2009) Are our brains structured to avoid refutations of belief in God? An experimental study. *Religion*, 39, 34–42.

Petsa, E. (2012). *A Study of Magical Thinking and its Relationship with Suggestibility in Greek Undergraduate Students* (MSc thesis). Lancaster University, UK.

Petty, R. E. & Cacioppo, J. T. (1986). *Communication and Persuasion: Central and Peripheral Routes to Attitude Change*. New York: Springer-Verlag.

Piaget, J. (1971/1929). *The Child's Conception of the World*. London: Routledge & Kegan Paul.

Pliner, P., Pelchat, M. & Grabski, M. (1993). Reduction of neophobia in humans by exposure to novel food. *Appetite*, 20, 111–123.

Principe, G. F. & Smith, E. (2008). Seeing things unseen: Fantasy beliefs and false reports. *Journal of Cognition and Development*, 9, 89–111.

Pronin, E., Wegner, D. M., McCarthy, K. & Rodriguez, S. (2006). Everyday magical powers: the role of apparent mental causation in the overestimation of personal influence. *Journal of Personality and Social Psychology*, 91, 218–231.

Radin, D. & Nelson, R. (2000). Meta-analysis of mind–matter interaction experiments: 1959 to 2000. www.spiritualscientific.com/yahoo_site_admin/assets/docs/Review_of_Mind-Matter_Interaction_Articles_19592000_RNG_Articles.12960830.pdf

Redfield, R. (1968). *The Folk Culture of Yucatan*. Chicago: University of Chicago Press.

Richards, C. A. & Sanderson, J. A. (1999). The role of imagination in facilitating deductive reasoning in 2-, 3-, and 4-year-olds. *Cognition*, 72, B1–B9.

Richert, R. A., Shawber, A. B., Hoffman, R. E. & Taylor, M. (2009). Learning from fantasy and real characters in preschool and kindergarten. *Journal of Cognition and Development*, 10, 41–66.

Richert, R. A. & Smith, E. I. (2011). Preschoolers' quarantining of fantasy stories. *Child Development*, 4, 1106–1119.

Robson, D. (2014). Cure for love: should we take anti-love drugs? *New Scientist*, February 15.

Rosenblatt, P. C. (1971). Communication in the practice of love magic. *Social Forces*, 49, 482–487.

Rosengren, K. S. & Hickling, A. (2000). Metamorphosis and magic: The development of children's thinking about possible events and plausible mechanisms. In K. S. Rosengren, C. Johnson, & P. Harris (Eds). *Imagining the Impossible: Magical, Scientific, and Religious Thinking in Children*. Cambridge: Cambridge University Press, pp. 75–98.

Rozin, P. Millman, L. & Nemeroff, C. (1986). Operation of the laws of sympathetic magic in disgust and other domains. *Journal of Personality and Social Psychology*, 4, 703–712.

Ruggiero, G. (1993). *Binding Passions*. Oxford: Oxford University Press.

Rumbaugh, D. M., Savage-Rumbaugh, E. S. & Sevcik, R. A. (1994). Behavioral roots of language: a comparative perspective of chimpanzee, child, and culture. In R. W.

Wrangham, W. C. McGrew, F. B. M. de Waal & P. R. Heltne (Eds), *Chimpanzee Cultures*. Cambridge: Harvard University Press, pp. 319–334.

Schooler, J. (2011). Unpublished results hide the decline effect. *Nature*, 470, 437.

Sejourne, L. (1976). *Burning Water: Thought and Religion in Ancient Mexico*. Berkeley: Shambala.

Selby, H. A. (1974). *Zapotec Deviance: The Convergence of Folk and Modern Sociology*. Austin: University of Texas Press.

Selten, J. P., Cantor-Craae, E. & Kahn, R.S. (2007). Migration and schizophrenia. *Current Opinion in Psychiatry*, 20, 111–115.

Shafran, R., Thordarson, M. A. & Rachman, S. (1996). Thought–action fusion in obsessive–compulsive disorder. *Journal of Anxiety Disorders*, 10, 379–391.

Sharon, T. & Woolley, J. D. (2004). Do monsters dream? Young children's understanding of the fantasy/reality distinction. *British Journal of Developmental Psychology*, 22, 293–310.

Sheldrake, R. (2013). *The Science Delusion*. London: Hodder & Stoughton.

Shu, S. B. & Carlson, K. A. (2014). When three charms but four alarms: identifying the optimal number of claims in persuasion settings. *Journal of Marketing*, 78, 127–139.

Siegel, R. K. (1977). Hallucinations. *Scientific American*, 237, 132–140.

Slater, A., Morison, V. & Rose, D. (1982). Visual memory at birth. *British Journal of Psychology*, 73, 519–525.

Smith, D. (2007). *Muses, Madmen, and Prophets: Rethinking the History, Science, and Meaning of Auditory Hallucination*. New York: Penguin Press.

Sobel, D. M. & Lillard, A. S. (2001). The impact of fantasy and action on young children's understanding of pretence. *British Journal of Developmental Psychology*, 19, 85–98.

Spengler, O. (1998). *Zakat Evropy*. Rostov: Feniks.

Stent, G. (1969). *The Coming of the Golden Age*. New York: Natural History Press.

Stratton, G. M. (1896). Some preliminary experiments on vision without inversion of retinal image. *Psychological Review*, 3, 611–617.

Subbotsky, E.V. (1983). Shaping moral action in children. *Soviet Psychology*, 1, 56–70.

Subbotsky, E. V. (1984). The moral development of the preschool child. *Soviet Psychology*, 22, 3–19.

Subbotsky, E. V. (1985). Preschool children's perception of unusual phenomena. *Soviet Psychology*, 3, 91–114.

Subbotsky, E. (1997). Understanding the distinction between sensations and physical properties of objects by children and adults. *International Journal of Behavioral Development*, 20, 321–347.

Subbotsky, E. (2000a). Phenomenlistic reality: the developmental perspective. *Developmental Review*, 20, 438–474.

Subbotsky, E. (2000b). Phenomenalistic perception and rational understanding in the mind of an individual: a fight for dominance. In K. S. Rosengren, C. N. Johnson & P. L. Harris (Eds), *Imagining the Impossible. Magical, Scientific and Religious Thinking in Children*. Cambridge: Cambridge University Press, pp. 35–74.

Subbotsky, E. (2001). Causal explanations of events by children and adults: can alternative causal modes coexist in one mind? *British Journal of Developmental Psychology*, 19, 23–46.

Subbotsky, E. (2004). Magical thinking in judgments of causation: can anomalous phenomena affect ontological causal beliefs in children and adults? *British Journal of Developmental Psychology*, 22, 123–152.

Subbotsky, E. (2005). The permanence of mental objects: testing magical thinking on perceived and imaginary realities. *Developmental Psychology*, 41, 301–318.

Subbotsky, E.V. (2006). *Stroyascheesia Soznaniye*. Moscow: Mysl.

Subbotsky, E. (2007). Children's and adult's reaction to magical and ordinary suggestion: are suggestability and magical thinking psychologically close relatives? *British Journal of Psychology*, 98, 547–574.

Subbotsky, E. (2009). Can magical intervention affect subjective experiences? Adults' reactions to magical suggestion. *British Journal of Psychology*, 100, 517–537.

Subbotsky, E. (2010a). Curiosity and exploratory behaviour towards possible and impossible events in children and adults. *British Journal of Psychology*, 101, 481–501.

Subbotsky, E. (2010b). *Magic and the Mind: Mechanisms, Functions and Development of Magical Thinking and Behavior.* New York: Oxford University Press.

Subbotsky, E.V. (2010c). Vyzhivaniye v mire machin. Vzgliad psyckhologa na prichiny very v sverchestestvennoye. *Natsional'niy Psickhologicheskiy Zhurnal*, 1, 42–47.

Subbotsky, E. (2011). The ghost in the machine: why and how the belief in magic survives in the rational mind. *Human Development*, 54, 126–143.

Subbotsky, E. (2012). Discrimination between fantastic and ordinary visual displays by children and adults. *The Open Behavioural Science Journal*, 2, 23–30.

Subbotsky, E. (2013) Sensing the future: reversed causality or a non-standard observer effect? *SAGE Open*, 6, 81–93.

Subbotsky, E. (2014). The belief in magic in the age of science. *SAGE Open*, 4, 1.

Subbotsky, E. (2015). Impossible phenomena as mediators in cognitive functioning and education. *SENTENTIA. European Journal of Humanities and Social Sciences*, 4, 120–137.

Subbotsky, E. (2016). Does magic exist? Presentation at the conference: Science of Magic. University of London, Goldsmith College. www.researchgate.net/publication/296959063_Does_Magic_Exist

Subbotsky, E., Hysted, C. & Jones, N. (2010). Watching films with magical content facilitates creativity in children. *Perceptual and Motor Skills*, 111, 261–277.

Subbotsky, E. & Matthews, J. (2011). Magical thinking and memory: distinctiveness effect for TV commercials with magical content. *Psychological Reports*, 109, 369–379.

Subbotsky, E. & Quinteros, G. (2002). Do cultural factors affect causal beliefs? Rational and magical thinking in Britain and Mexico. *British Journal of Psychology*, 93, 519–543.

Subbotsky, E. & Ryan, A. (2014). Motivation and belief in the paranormal in a remote viewing task. *The Open Behavioral Science Journal*, 8, 1–7.

Subbotsky, E. & Slater, E. (2011). Children's discrimination of fantastic vs realistic visual displays after watching a film with magical content. *Psychological Reports*, 112, 603–609.

Subbotsky, E. & Trommsdorff, G. (1992). Object permanence in adults: a cross-cultural perspective. *Psychologische Beitrage*, 34, 62–79.

Susskind, L. (2006). *The Cosmic Landscape: String Theory and the Illusion of Intelligent Design.* New York: Little, Brown and Co.

Szasz, T. (1971). *The Manufacture of Madness: A Comparative Study of the Inquisition and the Mental Health Movement.* London: Routledge & Kegan Paul.

Tambiah, S. J. (1990). *Magic, Science, Religion, and the Scope of Rationality.* Cambridge: Cambridge University Press.

Taylor, B. & Howell, R. (1973). The ability of three-, four-, and five-year-old children to distinguish ordinary from fantastic reality. *Journal of Genetic Psychology*, 122, 315–318.

Tegmark, M. (1998). The interpretation of quantum mechanics: many worlds or many words? *Fortschritte der Physik*, 46, 855–862.

Thalbourne, M. A. (1994). Belief in the paranormal and its relationship to schizophrenia-relevant measures: a confirmatory study. *British Journal of Clinical Psychology*, 33, 78–80.

Thomson, H. (2014). Cure for love: chemical cures for the lovesick. *New Scientist*, February 15, 26–27.

Tiger, L. & McGuire, M. (2010). *God's Brain.* New York: Prometheus Books.

Tipler, F. J. (1989). The Omega Point as *Eschaton*: answers to Pannenberg's questions for scientists. *Journal of Religion & Science*, 24, 217–253.

Tipler, F. (1994). *The Physics of Immortality: Modern Cosmology, God and the Resurrection of the Dead*. New York: Doubleday.

Tissot, R. & Burnand, Y. (1980). Aspects of cognitive activity in schizophrenia. *Psychological Medicine*, 10, 657–663.

Torrance, E. P. (1981). *Thinking Creatively in Action and Movement*. Bensenville: Scholastic Testing Service.

Trinkaus, E. (1993). Femoral neck-shaft angles of the Qafzeh-Skhul early modern humans, and activity levels among immature near Eastern Middle Paleolithic himinids. *Journal of Human Evolution*, 25, 393–416.

Tversky, A. & Kahneman, D. (1974). Judgement under uncertainty: heuristics and biases. *Science*, 185, 1124–1131.

Tylor, E. (1920/1871). *Primitive Culture*. New York: G. P. Putnam's Sons.

Volkova, P. (2013). *Most cherez bezdnu. Kniga 1*. Moscow: Izdatel'stvo.

Vyse, S. A. (2000). *Believing in Magic: The Psychology of Superstition*. New York: Oxford University Press.

Wegner, D. (2002). *The Illusion of Conscious Will*. Cambridge: MIT Press.

Wells, J. (2012). *Anti-Darwin*. Moscow: Al'pina Business Books.

Wertsch, J. V. (1991). *Voices of the Mind. Sociocultural Approach to Mediated Action*. Cambridge: Harvard University Press.

Wheeler, J. A. (1990). Information, physics, quantum: the search for links. In W. Zurek (Ed.), *Complexity, Entropy, and the Physics of Information*. Redwood City: Addison-Wesley, pp. 354–368.

Whiten, A. & Byrne, R. W. (1988). Tactical deception in primates. *Behavioral and Brain Sciences*, 11, 233–273.

Whitley, D. S. (2008). *Cave Paintings and the Human Spirit. The Origin of Creativity and Belief*. New York: Prometheus Books.

Wicker, A. W. (1969). Attitudes versus actions: the relationships of verbal and overt behavioral responses to attitude objects. *Journal of Social Issues*, 25, 41–78.

Wilson, C. (1971). *The Occult: A History*. London: Random House.

Winner, M. (2005). *Winner Takes All: A Life of Sorts*. London: Anova Books.

Woit, P. (2007). *Not Even Wrong: The Failure of String Theory and the Continuing Challenge to Unify the Laws of Physics*. London: Vintage.

Woolley, J. D. (2000). The development of beliefs about direct mental–physical causality in imagination, magic, and religion. In K. S. Rosengren, C. N. Johnson & P. L. Harris (Eds), *Imagining the Impossible: Magical, Scientific and Religious Thinking in Children*. Cambridge: Cambridge University Press, pp. 99–129.

Woolley, J. D., Boerger, E. A. & Markman, A. B. (2004). A visit from the Candy Witch: Factors influencing children's belief in a novel fantastic entity. *Developmental Science*, 7, 456–468.

Yang, M. & Roskos-Ewoldsen, D. R. (2007). The effectiveness of brand placement in the movies: levels of placements, explicit and implicit memory, and brand-choice behavior. *Journal of Communication*, 57, 469–489.

Zilboorg, G. & Henry, G. W. (1941). *A History of Medical Psychology*. New York: W. W. Norton & Co.

INDEX

Made in the USA
Columbia, SC
24 October 2024